# THEORETICAL ASPECTS OF Population Genetics

MOTOO KIMURA AND
TOMOKO OHTA

PRINCETON, NEW JERSEY
PRINCETON UNIVERSITY PRESS
1971

LC Card: 75-155963
ISBN: 0-691-08096-8 (hard cover edition)
ISBN: 0-691-08098-4 (paperback edition)
This book has been composed in Linofilm Baskerville
Printed in the United States of America
by Quinn & Boden Company, Inc., Rahway, N.J.

# THEORETICAL ASPECTS OF Population Genetics

MONOGRAPHS IN POPULATION BIOLOGY

EDITED BY ROBERT H. MacARTHUR

1. The Theory of Island Biogeography, by Robert H. MacArthur and Edward O. Wilson
2. Evolution in Changing Environments: Some Theoretical Explorations, by Richard Levins
3. The Adaptive Geometry of Trees, by Henry S. Horn
4. Theoretical Aspects of Population Genetics, by Motoo Kimura and Tomoko Ohta

TO JAMES F. CROW

# Preface

These essays were begun in the spring of 1969 when the senior author was at Princeton for a lecture series. The majority of the writing, however, has been a collaborative effort of the two of us in Mishima.

Throughout the book we have tried to show the importance of stochastic processes in the change of gene frequencies, and various topics ranging from molecular evolution to two-locus problems are discussed in terms of diffusion models. We have also attempted to come to grips with one of the most challenging problems in population genetics—the ways in which genetic variability is maintained in Mendelian populations.

It is pertinent to spend a few words on the development of mathematical theory in population genetics—a branch of genetics whose aim is to investigate the laws which govern the genetic structure of natural populations and, through that study, to clarify the mechanisms of evolution.

As a branch of science, its genealogy goes back to the theory of evolution by Darwin on the one hand and the theory of inheritance by Mendel on the other, synthesized by the method of biometry created by Karl Pearson, Francis Galton, and W. F. R. Weldon. That this synthesis was accomplished by the great triumvirate, R. A. Fisher, J. B. S. Haldane, and Sewall Wright is now well recognized, and we believe that it is one of the very great achievements in biology. The expanding rate at which papers in this field are now being published attests to the founder effect of these pioneering works.

It is remarkable that a systematic mathematical framework is built in a branch of biology that is similar to some branches of physics, especially statistical mechanics (and is beginning to be of comparable mathematical sophistication). Haldane, half a century ago, predicted that mathematical genetics would someday develop into a respectable branch of applied mathematics. His prediction has now been fully realized.

In these essays, however, we are interested in the mathematical theory, not as a branch of applied mathematics, but as an inevitable part of the theory of population genetics. So, whenever verbal description is practicable, we have preferred this to mathematical symbolism. We have attached a mathematical appendix for those who may be interested in a more rigorous approach. Most of the theory underlying these essays is given in the book by Crow and Kimura (1970), "An Introduction to Population Genetics Theory."

It was the physicist Ludwig Boltzman who said that there is nothing more practical than theory. We hope that some of the theories presented in these essays turn out to be useful in deepening our knowledge of the genetics of populations.

The senior author would like to use this opportunity to express his thanks to Drs. Robert H. MacArthur and Noboru Sueoka for their hospitality during his stay in Princeton.

We dedicate this book to James F. Crow, who has been a great teacher and inspiration in population genetics to us. He was kind enough to go over the original manuscripts of this book and gave many valuable suggestions.

We thank Miss Yuriko Matsumoto for typing the manuscripts and preparing the figures.

M. K.
T. O.

*National Institute of Genetics*
*Mishima, JAPAN*

# Contents

THEORETICAL ASPECTS OF

# Population Genetics

CHAPTER ONE

# Fate of an Individual Mutant Gene in a Finite Population

In natural populations, mutant genes arise in each generation, but the great majority of them are lost within a few generations. This is true not only for deleterious and selectively neutral mutations, but even for advantageous mutations, unless the advantage is enormous.

Only a small fraction of beneficial mutants are lucky enough to escape accidental loss in the first few generations of their existence. The fortunate few eventually attain a frequency high enough to protect them from further risk of chance extinction, so that their selective advantage can exert its effect and the mutant can become established in the population.

The fate of an individual mutant gene was first studied in detail by Fisher (1930a). He has shown, for example, that if a mutant gene is selectively neutral, the probability is about 0.79 that it will be lost from the population during the first seven generations. In an infinitely large population, eventually all the neutral mutants will be lost. On the other hand, if the mutant gene has a selective advantage of 1%, the probability of loss during the first seven generations is 0.78. As compared with the neutral mutant, this probability of extinction is less by only 0.01. However, in this case, the mutant gene will eventually spread to the entire population (assuming it is infinitely large) with a probability of about 0.02.

Fisher's 1930 paper treated these problems by a method which is now familiar in the treatment of "branching processes." This type of analysis needs much more sophisticated treatment than the deterministic treatment of gene frequency change developed by Haldane (1924–1931), whose results allow us to make such statements as "it takes about 1,000 generations until the gene frequency changes from 0.7% to 99.3% with selective advantage, $s = 0.01$." Such a deterministic treatment is still useful, but it cannot take into account the stochastic changes which are so important in treating the fate of individual mutant genes. Fisher's paper is remarkable in that as long ago as 1930 he could carry out a detailed analysis in this problem, although his method is limited to an infinite population. Yet, we believe that this is a masterpiece in the mathematical theory of population genetics.

Previously, Fisher (1922) had published a preliminary paper on the same subject, and this influenced Haldane (1927b) to work out the probability of fixation of mutant genes for the first time in a simple situation. For three decades following their work, not much progress was made in this field, though Wright's study on the distribution of gene frequencies under irreversible mutation contributed greatly to the later development of stochastic treatments in population genetics.

More than 30 years after the publication of Haldane's paper (1927b), we have finally begun to understand more about the fate of individual mutant genes in terms of the powerful diffusion methods based on the Kolmogorov forward and backward equations (cf. Kimura 1964). In particular, the average number of generations until extinction, and also, the time until fixation of an individual mutant gene in a finite population have been worked out (Kimura and Ohta 1969a, b). Likewise, the probability of fixation of mutant genes with arbitrary initial frequencies in a finite population has been computed (Kimura 1957, 1962 and 1964).

We shall first consider a selectively neutral mutant, since the results are simple and because this has an important bearing on the problem of molecular evolution. Throughout this chapter we shall denote by $p$ the initial frequency (relative proportion) of a mutant gene (allele) in the population of a diploid organism, and by $N$ and $N_e$ respectively the actual and effective population sizes (for $N_e$, see Chapter 3). We may note that for a single mutant gene which has just appeared in the population, $p = 1/(2N)$. We also note that $N$ and $N_e$ are in general different. Usually, the distribution of progeny number per individual deviates from Poisson distribution and this makes the ratio $N_e/N$ less than unity even if the population size remains constant and the numbers of breeding males and females are equal. According to Crow (1954), $N_e/N$ averages around 0.8 in man, but may vary considerably with different birth limitation patterns.

The frequency of the allele fluctuates from generation to generation, but for a neutral mutant allele, the proba-

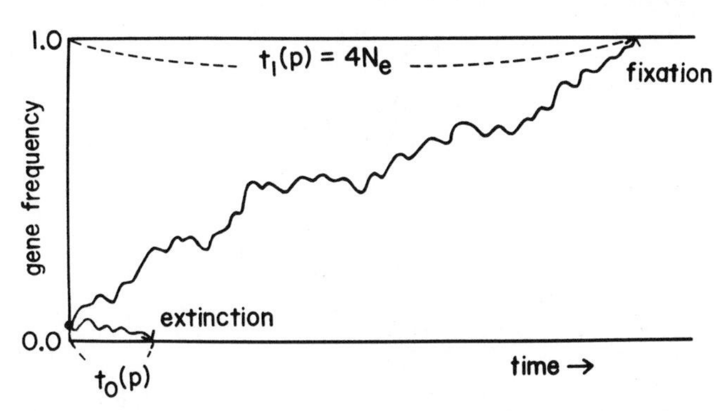


FIGURE 1.1. Diagram illustrating the courses of change in the frequencies of mutant genes, either leading to eventual fixation or to loss, starting from the initial frequency $p$. Abscissa: time; ordinate: mutant frequency in the population.

bility that the allele will eventually be fixed (established) in the population is $p$ (= initial frequency) and that of eventual loss is $1-p$. If the mutant allele is represented only once at the moment of its appearance, as is usually the case, $p = 1/(2N)$. Thus, a great majority $(1 - 1/2N)$ are eventually lost and only a tiny fraction $(1/2N)$ reach final fixation.

We denote by $t_0(p)$ the number of generations until loss of the mutant allele disregarding the cases of eventual fixation, and similarly, by $t_1(p)$, the number of generations until fixation disregarding the cases of eventual loss. Both are random variables and depend on the initial frequency, $p$. It can be shown that their means are

$$\overline{t_1(p)} = -\frac{1}{p}\{4N_e(1-p)\log_e(1-p)\} \tag{1}$$

and

$$\overline{t_0(p)} = -\frac{4N_e p}{1-p}\log_e p. \tag{2}$$

In particular, when $p = 1/(2N)$, we shall write $t_1$ and $t_0$ for $t_1(p)$ and $t_0(p)$. Then, it can be shown that sufficiently accurate approximations of their means are given by

$$\bar{t}_1 = 4N_e \tag{3}$$

$$\bar{t}_0 = 2\left(\frac{N_e}{N}\right)\log_e(2N) \tag{4}$$

with mean squares,

$$\overline{t_1^2} = 32N_e^2\left(\frac{\pi^2}{6} - 1\right) \tag{5}$$

$$\overline{t_0^2} = \frac{16N_e^2}{N} \tag{6}$$

assuming that both $N$ and $N_e$ are fairly large.

All these are approximations based on the diffusion models but they are in fairly good agreement with the results obtained by Monte Carlo experiments. The experiments were carried out using TOSBAC 3400, and some of the results for $\bar{t}_1$ are shown in Figure 1.2. The results confirm the prediction given in (3). Note that the time depends on only $N_e$.

Figure 1.3 shows the average time until extinction of a single mutant ($\bar{t}_0$) as a function of population number, assuming $N_e = N$. The solid line represents theoretical values and the dots give Monte Carlo results. They suggest a relatively large variance as compared with the mean. It can be shown that the variances of $t_1$ and $t_0$ are

$$\text{Var}\ (t_1) \approx 4.58N_e^2 \tag{7}$$

and

$$\text{Var}\ (t_0) \approx \frac{16N_e^2}{N} - \left[2\left(\frac{N_e}{N}\right)\log_e 2N\right]^2. \tag{8}$$

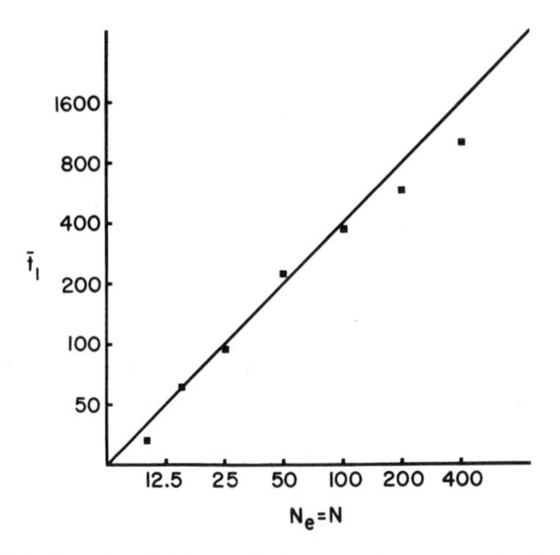


FIGURE 1.2. Results of Monte Carlo experiments to check formula (3) on the average number of generations until fixation for a neutral mutant. Experimental results are plotted as square dots, theoretical values are given by a line. Abscissa: effective population number; ordinate: time (in generations) until fixation. Each experimental value is the average of about 30 replicate trials excluding all cases of loss.

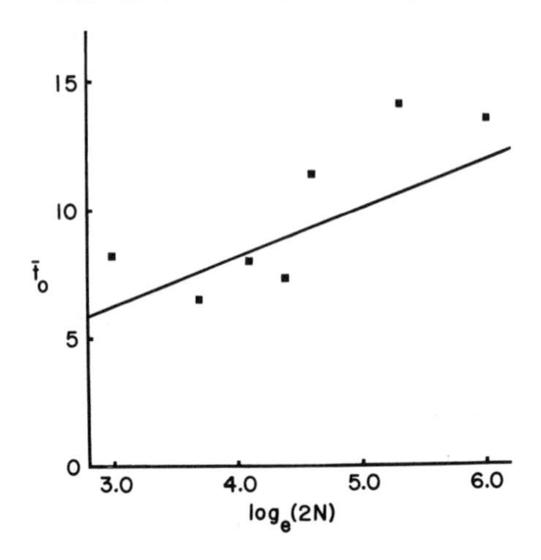


FIGURE 1.3. Results of Monte Carlo experiments to check formula (4) on the average number of generations until extinction ($\bar{t}_0$) of a neutral mutant. Experimental values are plotted as square dots, theoretical values are given by a line. In the experiments, actual and the effective numbers are assumed to be equal ($N = N_e$), and 195 ~ 200 replicate trials were made to obtain each experimental value.

For a very large $N$ and assuming $N = N_e$, we get $\sigma(t_0) \approx 4\sqrt{N}$. Thus, the standard deviation of the time until extinction is very large as compared with its mean, $\overline{t_0} \approx 2 \log_e 2N$. This comes from the fact that a neutral mutant occasionally may increase its frequency by chance, persisting in the population for quite a long time before its eventual loss. On the other hand, the time until fixation has a relatively small standard deviation of roughly half the mean, i.e. $\sigma(t_1) \approx (2.15)N_e$.

We can summarize the neutral case as follows. In a majority of cases (actually $1 - 1/2N$), the mutants are lost from the population but in a minority of cases ($1/2N$), they are eventually fixed, taking quite a long time. Figure 1.4 shows the probability distribution of the time until fixation obtained by Kimura (1970a). From the figure it may be seen

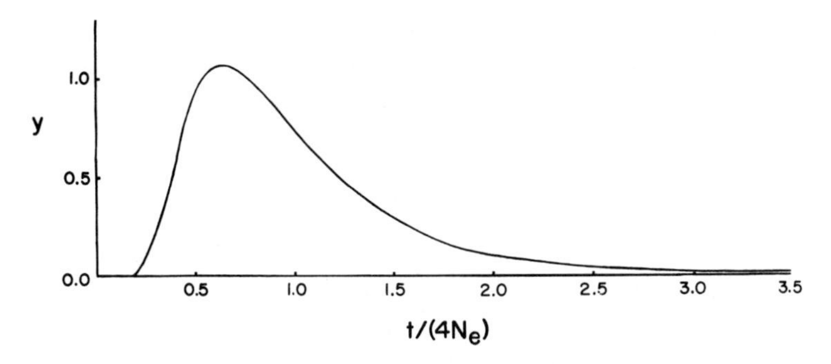


FIGURE 1.4. Probability distribution of the length of time until fixation of a selectively neutral mutant, excluding the cases of its eventual loss. Abscissa: time measured with $4N_e$ generation as the unit ($N_e$ = effective population number); ordinate: probability density.

that fixation before $(0.8)N_e$ generations is very unlikely to occur. Once again, Fisher's early insight was remarkable. In his 1930 book (p. 80) he says: "An inference of some interest is that in the absence of favourable selection, the number of individuals having a gene derived from a single mutation cannot greatly exceed the number of generations since its occurrence."

With natural selection, the problem becomes much more complicated, but as far as a pair of alleles at a single locus is concerned, we now have fairly good knowledge of the probability of fixation.

If we denote by $u(p)$ the probability that the mutant allele ultimately becomes fixed in the population, then,

$$u(p) = \frac{\int_0^p e^{-2SDx(1-x)-2Sx}\,dx}{\int_0^1 e^{-2SDx(1-x)-2Sx}\,dx} \tag{9}$$

where $S = N_e s$ and $D = 2h - 1$, in which $s$ and $sh$ are respectively the selective advantage of the mutant homozygote and heterozygote (Kimura 1957). If the mutant allele is semidominant (no dominant) such that $h = 1/2$, writing $s_1 = s/2$, the above formula reduces to

$$u(p) = \frac{1 - e^{-4N_e s_1 p}}{1 - e^{-4N_e s_1}}. \qquad (10)$$

Although based on the diffusion approximation, these formulae are very accurate unless $N_e$ is very small (cf. Ewens 1963, Carr and Nassar 1970). If the mutant gene is represented only once at the moment of its occurrence, we may put $p = 1/(2N)$. Writing $u$ for $u(1/2N)$, we have

$$u = \frac{1 - e^{-2(N_e/N)s_1}}{1 - e^{-4N_e s_1}}. \qquad (11)$$

For an advantageous mutant gene ($s_1 > 0$), this becomes approximately

$$u \approx 2s_1 \left(\frac{N_e}{N}\right), \qquad (12)$$

if $4N_e s_1 \gg 1$ and $2(N_e/N)s_1$ is small (Kimura 1964). When $N_e = N$, this reduces to the classical result obtained by Haldane (1927b) that the probability of ultimate fixation of an individual mutant gene is about twice its selective advantage ($u = 2s_1$). In man, $N_e/N$ is around 2/3 so that this probability must be modified to give $u = (4/3)s_1$. This agrees with the result obtained by Kojima and Kelleher (1962) who used the branching process method.

Formula (12) shows that for slightly advantageous new mutants only a tiny minority are lucky enough to spread into the entire species, while the remaining majority are lost by chance even if they have a definite selective advantage. This means, for example, that in a population in which $N_e = N$, for every mutant gene having selective advantage of 0.5% that becomes fixed in the population, 99 equally advantageous mutants have been lost, without ever being used in evolution.

The fact that the majority of mutants, including those having a slight advantage, are lost by chance is important in considering the problems of evolution by mutation, since the overwhelming majority of advantageous mutations are likely to have only a slightly advantageous effect. Note that a majority of mutations with large effect are likely to be deleterious. Fisher (1930b) emphasized that the larger the effect of the mutant, the less is its chance of being beneficial.

In our opinion, this fact has not fully been acknowledged in many discussions of evolution. It is often tacitly assumed that every advantageous mutation that appears in the population is inevitably incorporated. Also, it is not generally recognized that this fact can set an upper limit to the speed of adaptive evolution, because the frequency of occurrence of advantageous mutations must be much lower than that of deleterious mutations. (Incidentally, this fact gives a great evolutionary advantage to a large population.)

In order to see this, let $K$ be the rate of gene substitution in evolution measured by the average number of gene substitutions per generation. Then, if $\nu_m$ is the number of advantageous mutant genes which appear in the entire population each generation,

$$K = \nu_m u = 2s_1 \left(\frac{N_e}{N}\right) \nu_m. \qquad (13)$$

Note that each gene substitution takes a fairly large number of generations for this to occur. The mutant allele must increase from a very low frequency to a very high frequency and finally to fixation. As mentioned earlier, it takes on the average $4N_e$ generations for a neutral mutant to reach fixation. For definitely advantageous genes, the time is somewhat less. However, if we wait long enough and consider averages, then formula (13) holds. If we denote by $v$ the mutation rate per gamete per generation for advantageous mutations, such that $v = \nu_m/(2N)$, then (13) becomes

$$K = 4N_e s_1 v, \tag{14}$$

where $4N_e s_1 \gg 1$ for the formula to be valid. For example, in a population of effective size $N_e = 10^4$, if the selective advantage of the mutant is 1% ($s_1 = 10^{-2}$), we have $K = 400v$. Although the value of $v$ is unknown, considering that the lethal mutation rate per gamete per generation is some 1.5% in *Drosophila*, it is unlikely that $v$ is larger than $10^{-3}$.

As long as we accept the Darwinian view of adaptive evolution by natural selection and the random nature of gene mutations as disclosed by Mendelian genetics, we must assume that advantageous mutations cannot occur nearly as frequently as deleterious or even neutral mutations. Otherwise, the selectionist theory of evolution is useless and we must rely on some other theory. If we assume that advantageous genes occur by mutation only 1/1,000 as often as lethal genes, i.e. $v = 1.5 \times 10^{-5}$, and still assuming $N_e = 10^4$ and $s_1 = 10^{-2}$, then we have $K = 400 \times 1.5 \times 10^{-5} = 6 \times 10^{-3}$ or about one substitution every 170 generations. If the selective advantage is 1/10 as large ($s_1 = 10^{-3}$), $K$ becomes $6 \times 10^{-4}$. In general, we may conclude that unless $N_e$ is of the order of a million or more, it is unlikely that $K = 1$ is attained with $s_1 = 10^{-3}$ or less.

It is evident from these considerations that the formula for the probability of gene fixation has important applications in evolutionary theory. It might be rather unexpected, however, that it has also an application in the theory of plant and animal breeding. For many years following the work of Smith (1936) and Hazel (1943) on selection indices, it had been almost the sole concern in the theory of artificial selection to maximize the genetic gain (usually denoted by $\Delta G$) per generation. However, the problem of the selection limit, that is, the ultimate genetic improvement that may be attained under a given scheme of artificial selection for a given population, was largely left out of consideration. It was Robertson (1960) who pioneered in the development of a theory of selection limits based on the theory of gene fixation in a finite population. In the simplest case of an additive quantitative character, he showed that the ultimate gain is $2N_e$ times the gain in the first generation. This can be seen as follows. From (10), assuming that $4N_es_1$ is small, we obtain

$$u(p) = p + 2N_es_1p(1-p) + \cdots . \qquad (15)$$

Since $u(p) - p$ is the ultimate gain expected by continued selection when the attainable limit is unity, this formula shows that the ultimate gain is approximately equal to $2N_e$ times $s_1p(1-p)$, the gain in the first generation. Robertson also showed that the highest limit is attained when the better 50% of the individuals are saved for breeding in each generation. This conclusion was also obtained earlier by Dempster (1955).

Since the work of Robertson (1960), a number of papers have been published in this field (see, for example, Hill 1969). Recently, the effect of linkage on the selection limit has been clarified (Hill and Robertson 1966), and also the theory has been extended to cover the multiple locus situation (Robertson 1970b).

Next, we shall turn to deleterious mutations and consider their fate in a finite population, especially when the product of selection coefficient and effective population number is large (cf. Kimura and Ohta 1969b). In this case, as shown before, the mutant gene will almost certainly be eliminated before it reaches any appreciable frequency. There is virtually no probability of its becoming established in the population.

Let $s'$ and $s'h$ be the selection coefficient against mutant homozygote and heterozygote respectively. If the mutant gene is semidominant ($h = 1/2$) and if it is sufficiently deleterious that $2N_e s' \gg 1$, then it can be shown that the average number of generations until extinction of an individual mutant gene is

$$\overline{t_0} = 2\left(\frac{N_e}{N}\right)\{\log_e(2N) - \log_e(2N_e s') + 1 - \gamma\}, \quad (16)$$

where $\gamma$ is Euler's constant, 0.577. Figure 1.5 shows the result of Monte Carlo experiments carried out to check this formula. In these experiments, $2N = 150$, $2N_e = 100$, and for each value of the selection coefficient, 50 trials were made except for the case of $s' = 0$ for which 500 replications were made.

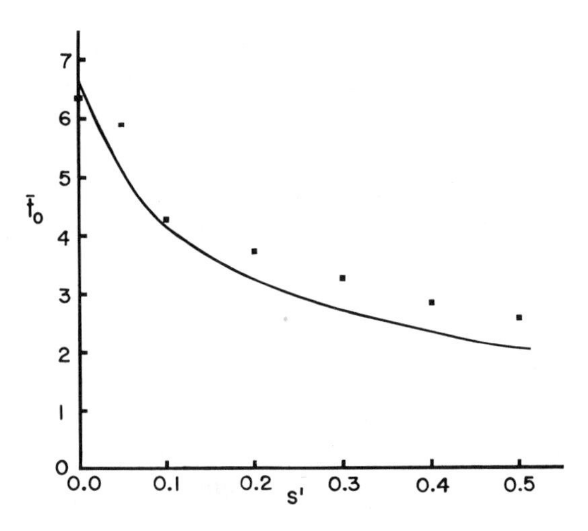


FIGURE 1.5. Results of Monte Carlo experiments on the average number of generations until extinction of a semidominant deleterious mutant. Experimental results are plotted as square dots, and theoretical values computed by formula (16) are given by a curve. Abscissa: selection coefficient against the mutant; ordinate: number of generations until extinction. (From Kimura and Ohta, 1969b.)

If the mutant gene is deleterious, but completely recessive, assuming that $2N_e s' \gg 1$, we obtain approximately,

$$\bar{t}_0 = 2\left(\frac{N_e}{N}\right)\left\{\log_e(2N) - \frac{1}{2}\log_e(2N_e s') + 1 - \frac{\gamma}{2}\right\}. \quad (17)$$

Table 1.1 shows the values of $\bar{t}_0$ for $2N$ from $10^3$ to $10^6$ assuming $N_e/N = 0.5$ and $s' = 0.01$ for the deleterious mutant. The table reveals the interesting fact that for most of the realistic range of population sizes for human populations, the time until extinction does not differ very much for mildly deleterious and neutral mutant genes.

TABLE 1.1. The average number of generations until extinction for neutral, semidominant deleterious and recessive deleterious mutants, computed by using formulae (4), (16) and (17), for five levels of population number. The effective population number ($N_e$) is assumed to be half as large as the actual population number ($N$). Also, a selection coefficient $s' = 0.01$ is assumed for deleterious mutants.

| $2N$ | Neutral | Semidominant deleterious | Recessive deleterious |
|---|---|---|---|
| $10^3$ | 6.9 | 5.7 | 6.8 |
| $10^4$ | 9.2 | 5.7 | 8.0 |
| $10^5$ | 11.5 | 5.7 | 9.1 |
| $10^6$ | 13.8 | 5.7 | 10.3 |
| $10^{12}$ | 27.6 | 5.7 | 17.2 |

CHAPTER TWO

# Population Genetics and Molecular Evolution

From the standpoint of population genetics, the process of evolution consists of a series of gene substitutions. For each gene substitution, a mutant allele, which is initially rare, increases its frequency and spreads into the entire population (species), finally reaching the state of fixation. Each of these events must usually take a large number of generations. The question that naturally arises is how many gene substitutions are required to transform one species to another, one genus to another, etc. Also, we would like to know at what rate the genes are being substituted in the process of evolution.

Despite the impressive amount of studies that had been made since the days of Darwin about evolution at the phenotypic level, no definite clue to answer these questions had been obtained until we started to understand genes in terms of molecular structure. (For a lucid discussion of this subject, the reader is invited to read Crow's 1969 article on molecular evolution and population genetics presented in the Twelfth International Congress of Genetics.)

Before we consider the problem of molecular evolution from the standpoint of population genetics, we would like to mention some important early work by Haldane (1949) relating the rate of evolution and gene substitution. Based on fossil records, he studied the rate of change in quantitative measurements such as body size and tooth length and concluded that evolution is generally a very slow process. In a typical case, such as the evolution of horse, the tooth length changed at the rate of a few percent per million years. He coined the word *darwin* to represent change at the rate of $10^{-6}$ per year. In terms of this, the typical rate of evolution in quantitative characters is about one-tenth of a darwin. He then conjectured that, at each genetic locus, alleles are substituted at an interval of a few million generations. Later, based on his conception of "the cost of natural selection" (Haldane 1957, see Chapter 5), he came to the conclusion that in horotelic (standard rate) evolution, genes are substituted in the population at the rate of about once in every 300 generations (per genome) on the average.

In order to obtain a reliable estimate of the rate of gene substitutions in evolution, we must have a way by which genes in remotely related organisms such as carp and man can be compared structurally together with fairly reliable information on the time since their divergence. Recent development in comparative studies of amino acid sequence in proteins and the modern method of isotope dating of fossils have fulfilled such requirements. Thus, information obtained from molecular evolution may be used to study the problem of gene substitution in population genetics (Kimura 1968a, 1969b). A similar attempt has also been made by King and Jukes (1969).

Among various proteins, hemoglobins seem to have been studied most extensively, and there is a great amount of data (cf. Dayhoff 1969) from which the rate of evolution of hemoglobin chains may be computed.

The method of calculation is as follows: Consider two homologous polypeptides and let $n_{aa}$ be the number of amino acid sites for which comparison of amino acids can be made. Since we are concerned with amino acid substitution only, we exclude deletions and insertions if they exist between the two polypeptide chains.

This does not mean that we consider the latter changes unimportant. Structural changes, especially duplication of chromosome segments, must have played a very important role in progressive evolution, allowing one of the duplicated segments to accumulate mutations and acquire a new function while the other segment retains the old function necessary to survive through the transitional period. Also, the old function may be retained permanently, thereby permitting the organism to increase in genetic complexity. For a stimulating discussion, see Ohno (1970).

In addition, amino acid deletions and insertions give very important information relating to phylogeny. In the present chapter, however, we are mainly concerned with amino acid substitutions and will disregard such structural changes.

Let $d_{aa}$ be the number of sites in which the two chains are different. For each site compared, a difference in amino acids means there has been at least one substitution in the course of evolution. In the literature of molecular evolution, differences of amino acids between homologous proteins are often referred to simply as the result of randomly distributed point mutations, but from our standpoint, they are the result of gene substitutions. So, if we denote the mean number of substitutions per amino acid site over the whole period of evolution by $K_{aa}$, then, assuming independence of substitutions among different sites, the probability of no substitution at any site is $e^{-K_{aa}}$ by the Poisson law, and therefore the probability of at least one substitution is $1 - e^{-K_{aa}}$. Thus,

$$d_{aa} = n_{aa}(1 - e^{-K_{aa}}). \qquad (1)$$

If we denote by $p_d$ the proportion of differentiated sites such that $p_d = d_{aa}/n_{aa}$, then we obtain from (1), the estimator

$$K_{aa} = -\log_e (1 - p_d). \qquad (2)$$

This procedure, including the Poisson correction, was first

used by Zuckerkandl and Pauling (1965). The standard error of $K_{aa}$ may be computed from

$$\sigma_K = \sqrt{\frac{p_d}{(1 - p_d)n_{aa}}} \tag{3}$$

(Kimura 1969a).

The rate of substitution per amino acid site per year may then be obtained by

$$k_{aa} = \frac{K_{aa}}{2T}, \tag{4}$$

where $T$ is the number of years that have elapsed since the evolutionary divergence of the two chains from their common ancestor.

Figure 2.1 illustrates the comparison of the hemoglobin $\alpha$ chain of man with that of the carp. By comparing these two chains, we find that they have the same amino acid at 72 sites and different amino acids at 68 sites. In addition, these two chains are differentiated by insertions or deletions that amount to three amino acids. So, excluding those unmatched positions, we have $d_{aa} = 68$ and $n_{aa} = 68 + 72$

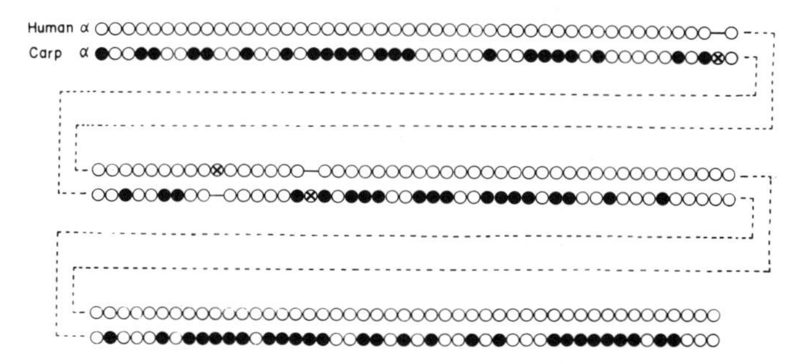


FIGURE 2.1. A diagram showing the number and positions of amino acid differences between human and carp hemoglobin $\alpha$ chains. A difference due to amino acid substitution is indicated by a solid circle, while a difference due to deletion or insertion is indicated by a crossed circle or a bar.

= 140. From formulae (2) and (3), we get $K_{aa} = 0.665$ and $\sigma_K = 0.082$. As to the length of time since divergence ($T$), we are reasonably sure that the common ancestor of man and carp lived in the Devonian period, which is known as the age of fishes and which dates back to about 350 to 400 million years (Figure 2.2). Thus, taking $2T = 750 \times 10^6$, we obtain the rate of substitution $k_{aa} = 8.9 \times 10^{-10}$ per amino acid site per year. Similar calculations may be made by comparing the $\alpha$ chain of man with horse, cow, pig, rabbit and sheep $\alpha$ chains. This gives, as the average of five values, $K_{aa} = 0.141 \pm 0.014$. Since the mammals probably diverged from their common ancestor some 80 million years ago, we may take $2T = 160 \times 10^6$, giving the rate of substitution $k_{aa} = (8.8 \pm 0.9) \times 10^{-10}$ per amino acid site

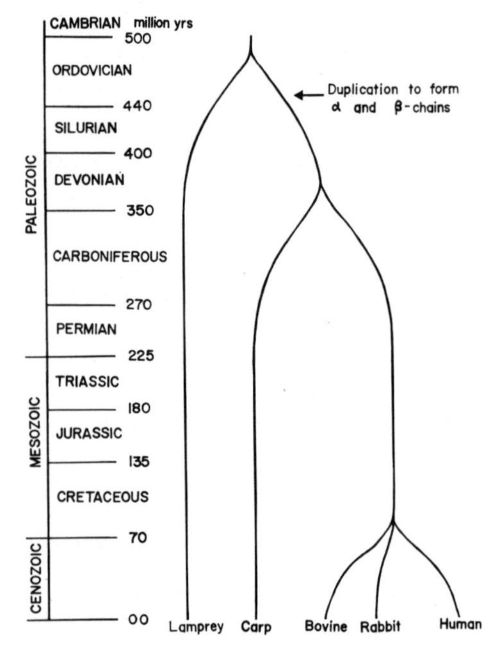


FIGURE 2.2. A phylogenetic tree for vertebrate evolution. (From Kimura 1969b.)

per year, which is very near to the value obtained by comparing the $\alpha$ chains of man and the carp. We can also compare the $\alpha$ chain of the mouse with those of the human, horse, cow, pig, rabbit and sheep, and these comparisons give $k_{aa} = (10.9 \pm 0.9) \times 10^{-10}$ per amino acid site per year, which is again very near to the values obtained above. It is interesting to note here that the generation time in the two lineages, one leading to man and the other leading to the mouse, must differ a great deal, and yet the rates of amino acid substitutions as estimated above are almost the same per year. By making a similar analysis with respect to $\beta$ chains, the same tendency was discovered.

Comparison of the $\alpha$ and $\beta$ chains in various animals leads to even more interesting conclusions. It is generally accepted that these chains originated by a gene duplication in the remote past. If we compare the $\beta$ chain of man with the $\alpha$ chains of human, mouse, rabbit, horse, cow and carp, we find that these $\alpha$ chains differ from the $\beta$ chain of man almost the same extent. For example, for the comparison of human $\beta$ vs. human $\alpha$, excluding deletions or insertions that amount to nine amino acids, we have $K_{aa} = 0.776$, while for the comparison of human $\beta$ vs. carp $\alpha$, we have $K_{aa} = 0.807$. It can be shown that the standard error for the difference of these two estimates of $K_{aa}$ is about 0.091, and therefore the two observed $K_{aa}$ values agree within the limit of statistical error.

The same is true for the other comparisons between $\alpha$ and $\beta$ chains. These results indicate that the two structural genes corresponding to the $\alpha$ and $\beta$ chains, after their separation by duplication, have diverged from each other independently and to the same extent on whatever evolutionary line they are placed; the amount of divergence is the same irrespective of whether the compared $\alpha$ and $\beta$ chains are taken from the same organism (man) or from man and carp, which have evolved independently over 350 million years. As to the time of gene duplication, it is reasonable to assume that it occurred at the jawless stage of vertebrate evolution, for the globin found in lamprey blood is a monomer. So if we assume that the formation of $\alpha$ and $\beta$ chains goes back to the end of Ordovician period (about 450 million years ago), we obtain $k_{aa} = 8.9 \times 10^{-10}$ per amino acid site per year, which is very similar to the previous values.

Summing up, the hemoglobin molecules have undergone amino acid substitutions in the past 450 million years throughout the diverse lines of vertebrate evolution at a remarkably constant rate of about $10^{-9}$ per amino acid site per year. Also, the above analysis suggests that the substitutions occurred fortuitously, for it is hard to imagine a pattern of selection that would lead to such results.

A similar constancy in the rate of amino acid substitution is apparent in Figure 2.3, in which comparison is made of cytochrome c from diverse organisms.

To the eyes of evolutionists trained in the Darwinian theory of natural selection, it must appear extraordinary that the rate of substitution depends only on time measured in years and is almost independent of generation time, living conditions or even the genetic background. Such remarkable constancy per year is most easily understood by assuming that in diverse lines the rate of production of neutral mutations per individual per year is constant. This inference is based on the simple principle (Kimura 1968a) that for neutral mutations, the rate of gene substitution in a population is equal to the production of new mutations per gamete because, for such a mutation, the probability of gene fixation is equal to the initial frequency. It may be rather surprising that the rate of gene substitution for neutral mutations does not depend on population size. This may be explained most simply as follows. In a population of actual size $N$, the probability of eventual fixation of an

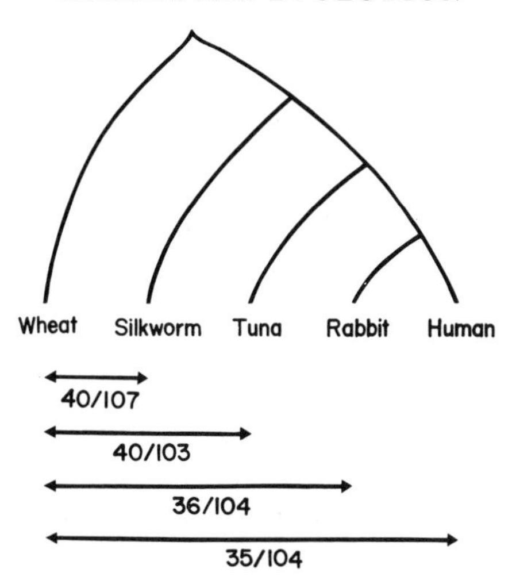


FIGURE 2.3. The fraction of amino acid differences in cytochrome c molecules of various animals as compared with that of wheat. In counting the number of different sites, unmatching parts due to insertion or deletion were excluded.

individual mutant gene that is selectively neutral is $1/(2N)$. If $v_0$ is the mutation rate per gamete per unit time for neutral mutations, then the number of newly arisen neutral mutants in the entire population is $2Nv_0$ of which fraction $1/(2N)$ are lucky enough to reach eventual fixation. So the rate of gene substitution per unit time is $K = 2Nv_0 \times 1/(2N) = v_0$, which is equal to the rate of production of neutral mutants per gamete per unit time. Note that, as explained in Chapter 1, for each neutral mutant destined to reach fixation, it takes a number of generations equal to about four times the effective population size to reach fixation; however, if we wait long enough and take an average of the number of substitutions per unit time, we get $K = v_0$.

On the other hand, if mutant substitution is due to natural selection acting on advantageous mutations, as shown in the previous chapter (formula 14), the rate of gene substitution depends not only on the rate of their production ($v$) but also on effective population number ($N_e$), and their selective advantage ($s_1$), and it is difficult to believe that in natural populations, these are always adjusted such that their product is constant.

The same conclusion was arrived at by King and Jukes (1969), who measured the rate of evolution for seven proteins. According to their estimates, the rate of substitution ranges from $0.33 \times 10^{-9}$ (insulins) to $4.29 \times 10^{-9}$ (fibrinopeptide A). Also, it is known that histones have an extremely slow rate of change amounting to two changes in approximately 1.5 billion years per 100 amino acid sites (cf. Dayhoff 1969), but this appears to be an exception. To describe the rate of molecular evolution, it might be convenient to use a unit, the *pauling*, defined as the rate of substitution of $10^{-9}$ per amino acid site per year (Kimura 1969a). In terms of this, the hemoglobin rate is nearly one pauling, while the rates for seven proteins computed by King and Jukes range from 33 centipaulings (insulin) to 4.3 paulings (fibrinopeptide A), with the average rate 1.6 paulings ($\bar{k}_{aa} = 1.6 \times 10^{-9}$ per amino acid site per year). This means that in each codon at a mammalian genome, the average interval between substitutions is roughly half a billion years.

As to the differences in rates of evolution among various proteins, the most plausible interpretation is that the different fractions of amino acid substitutions are neutral, as pointed out by King and Jukes. In this connection, an important result was recently obtained by Fitch and Markowitz (1970), whose analysis of cytochrome c suggests that in this protein only about 10% of amino acid sites are changing at any moment in the course of evolution, while in fibrinopeptide A, most of the amino acids are evolving.

If we assume that the average rate of amino acid substitution is 1.6 paulings and note that about 20% of base substitutions are synonymous, this leads to an estimate of ⅓

$\times$ 1.6 $\div$ 0.8, or about $2/3$ pauling per nucleotide. In man the total number of nucleotide pairs making up the haploid chromosome set is estimated to be about $3 \sim 4 \times 10^9$ (cf. Muller 1958, Vogel 1964). If evolution occurs at the same rate for total DNA as it does for those nucleotides which code for the observed amino acid changes, then the total rate is $2 \sim 3$ nucleotide substitutions per year. This is some four or five times higher than the corresponding figure estimated earlier (Kimura 1968a). In the line leading to man, the average generation time was probably longer than 10 years. This gives a rate of nucleotide substitution per generation of at least 20, making the contrast still greater with Haldane's (1957) estimate of $1/300$ per generation as the standard rate of gene substitution in evolution. Considering the amount of selective elimination that accompanies the process of gene substitution (substitutional load, see Chapter 5), the most natural interpretation is, we believe, that a majority of *molecular mutations* that participate in evolution are almost neutral in natural selection.

The above method of extrapolating the rate of mutant substitution from an amino acid site to the whole genome assuming some $4 \times 10^9$ nucleotide pairs might be criticized on the grounds that the chromosome of higher organisms is a multistrand structure so that our calculations overestimate the true genomic rate by a factor equal to the number of strands. The same criticism applies if many non-heritable copies of a gene are produced in each generation on the chromosome (cf. Callan 1967), or if not all the DNA is genetic.

The *Drosophila* data on this point are inconclusive. According to Chovnick's (1966) analysis of the *rosy* cistron in *Drosophila melanogaster,* the cistron covers $8.86 \times 10^{-3}$ crossover units on the third chromosome. Since the DNA content per nucleus is about $1/20$ the amount in man (UN Report 1958), the number of nucleotide pairs per haploid genome is about $2 \times 10^8$. The third chromosome is about 100 crossover units long and corresponds roughly to $2/5$ of the genome. Thus the number of nucleotide pairs making up the *rosy* cistron is estimated to be

$$2 \times 10^8 \times (2/5)10^{-2} \times 8.86 \times 10^{-3} \doteqdot 7 \times 10^3,$$

i.e. about 7,000. This cistron is concerned with the production of the enzyme xanthine dehydrogenase (cf. Chovnick 1966). According to Shinoda (1967), the molecular weight of this enzyme is about 260,000. Assuming that the average molecular weight of an amino acid is 130, the total number of amino acids making up the xanthine dehydrogenase is some 2,000. This means that the structural gene should consist of about 6,000 nucleotide pairs, which agrees very well with the 7,000 estimated above. A similar calculation was made by Chovnick et al. (1962). This would argue against the possibility that the chromosome consists of a number of identical DNA strands, though a possibility cannot be excluded that xanthine dehydrogenase is a dimer and the structural genes consist of only 3,000 nucleotide pairs, in which case the chromosome consists of two identical DNA strands (each with a double helical structure).

On the other hand the *ma*-1 region of the *X* chromosome has 34 cistrons so far recognized in a map distance of about 2.3 crossover units (Lifschytz and Falk 1969), so this would suggest that the *rosy* locus is exceptionally large or that its recombination is unusual. If the *ma*-1 region is typical, then the estimated number of genes is 10 times as large or the size of the cistron is $1/10$ as large.

Finally, at the opposite extreme, if we take the total number of salivary bands as about $10^4$ and, somewhat naively, equate this to the number of genes, and take the typical size of a gene to be 500 nucleotides, this leads to an estimate of $5 \times 10^6$ nucleotide pairs. This is only about $1/40$ the actual amount of DNA in the haploid genome, $2 \times 10^8$ as mentioned above. This would argue for polyteny, although Schalet's (1969) study of the bobbed locus offers evidence against the "master-slave" hypothesis.

So, for the moment, the question of whether the number of genes is considerably larger than previously estimated, or whether the size of a gene is very large, or whether much of the DNA is redundant or non-informational must be left in abeyance (but, see Ohta and Kimura, 1971c). For this reason, when we write of molecular mutations we mean all changes in DNA, whatever the exact relation of DNA to the chromosome and the number of genes turns out to be.

It might also be argued that the load can be considerably reduced if we assume a very small (yet large enough to be effective) selective advantage based on competition (cf. Maynard Smith 1968a). However, this will violate another principle, as mentioned already in Chapter 1, that advantageous mutations cannot occur at a rate comparable to that of deleterious mutations.

As given in formula (14) of Chapter 1, for gene substitution by natural selection,

$$K = 4N_e s_1 v. \tag{5}$$

From the estimation made so far, $K$ is at least 20. The effective population size, $N_e$, of most primates should not be very large. According to Deevey (1960), at the inception of the hominid line one million years ago, the total population number was probably about 125,000. Considering population fluctuations and inequality in the number of breeding males and females, it is probable that the effective population number is at most $10^5$, and perhaps very much less. So we may take $N_e = 10^5$. Then, if we assume that the selection coefficient is 0.1%, i.e. $s_1 = 10^{-3}$, we obtain from (5), $v = 20/400 = 0.05$. This means that in each generation one out of twenty gametes carries a newly arisen advantageous mutation. In other words, advantageous mutations occur just as frequently as recessive lethal mutations. So, we cannot assume a very small selective advantage.

If we accept that a majority of evolutionary alterations of amino acids were caused by random fixation of neutral mutations in the population, then we must assume that neutral molecular mutations are occurring at a rate high enough to account for the rate of amino acid substitution in evolution. In man, this should amount to 2 or 2.5 per gamete per year, and with the present generation time of roughly 30 years, it amounts to 60 ~ 75 neutral molecular mutations per generation.

On the other hand, the frequency of deleterious mutations must be much lower, probably less than one per individual per generation; otherwise, even with reinforcing type epistasis between deleterious genes (Kimura and Maruyama 1966), the mutational load must be unbearably high for human populations. Although we do not know at present what fraction of the total DNA in the genomes of higher organisms codes for protein and therefore should be considered as genetic material in the conventional sense, we believe the conclusion is inevitable that in the human genome, nucleotide substitution has an appreciable effect on fitness in only a small fraction of DNA sites, possibly less than 1% of the total. It is probable that natural selection operating to reduce the mutational load may have done so more by changing the harmfulness of mutations than by reducing the rate of nucleotide error. Structural changes may be capable of repair, as in repair of UV-induced thymine dimers. Natural selection would place a large value on proteins which could function equally well with amino acid replacements. Even for bacterial DNA, a large fraction of nucleotide substitutions appear to have very little effect in fitness, as indicated by the remarkable experiment by Cox and Yanofsky (1967), who used the Treffers' mutator gene in *E. coli.* This gene causes preferentially the transversion AT → CG in the genome of *E. coli.* According to them the estimated rate of mutation is $3.5 \times 10^{-6}$ per A-T pair per generation. With $3.7 \times 10^{6}$ nucleotide pairs making up the genome of *E. coli,* and with G-C content of 50%, this amounts to some seven molecular mutations per bacterium per generation. In a strain containing this gene, they observed an increase of 0.2–0.5% in the G-C composition of the DNA after 80 subcultures, which corresponds to 1,200–1,600 cell generations. On the other hand, the expected increase in G-C composition based on the above mutation rate and assuming 50% G-C in the genome of *E. coli* is about 0.21–0.28%, agreeing with the observed increase in G-C content. Such agreement suggests that the majority of base substitutions are nearly neutral and incorporated into the gene pool of the bacteria population (note also that bacteria cannot tolerate more than 50% mortality per division, as pointed out by Crow 1969). This is consistent with their observation that the strain is fully viable after accumulation of more than 7,000 base substitutions in the course of the experiments. Also, the results of competition experiments between the mutator strain and coisogenic normal strain using a chemostat support the view that most of the substitutions are almost neutral (Gibson, Scheppe and Cox 1970).

Recently the way of measuring directly the rate of DNA base substitution in evolution has become available by the DNA hybridization technique. According to Laird et al. (1969), the estimated rate of DNA evolution among artiodactyls as measured by such a technique is faster than hemoglobin, but only slightly slower than fibrinopeptide; so the above estimate of 1.6 paulings appears to apply roughly to the entire genome, not just to the portion of genome coding for protein.

Laird's results with respect to rodents are less consistent, suggesting a very rapid evolutionary rate in their early stage of divergence. Namely, the estimated number of nucleotide substitutions between mouse and rat or between mouse and hamster is roughly the same as that between mouse and guinea pig. If the mouse-rat divergence is assumed to be only $^1/_{10}$ as old as the mouse-guinea pig divergence, as Laird assumed, his data suggest that the evolutionary rate obtained from the former comparison is some 10 times faster than the latter, which has about the same rate as artiodactyls. If this turns out to be correct, it would contradict our thesis that the rate of molecular evolution is uniform among different lines. There is a possibility, however, that these divergences are all equally old, so the rates of molecular evolution per year are the same among rodents and these rates are comparable to those among artiodactyls.

At this point, it is pertinent to mention the pioneering work of Sueoka (1961, 1962), who has contributed probably more than anybody else to convincing biologists that G-C content is an important parameter in considering evolution and phylogeny. In his theoretical study, Sueoka used the term "effective base conversion rates" to denote the rate of nucleotide substitutions that change an $\alpha$ (A-T or T-A) pair to a $\gamma$ (G-C or C-G) pair in the population. If the majority of base substitutions in evolutions are selectively neutral, as we have endeavored to show, his "effective base conversion rates" are indeed equal to corresponding mutation rates (cf. Freese 1962).

The above arguments do not mean that a majority of nuclear DNA is functionless. They should indicate that the human genome, and those of higher organisms in general, are endowed with a buffer mechanism against the disturbing effect of molecular mutations.

The unique structure of DNA that makes it so useful as

the chemical basis of heredity—namely, the fact that the base composition can change without alteration of the chemical, structural and replicative properties of the molecule—also means that if DNA has other functions in the cell (such as providing a structure of some sort) it can probably carry out these functions just as well with changes in composition. The chemical properties and structure of DNA are invariant with respect to substitutions of nucleotides.

Furthermore, it is probable that even if most mutants were neutral at the time of their substitution in the population, later a small fraction of them may have become essential to the organism, through ensuing establishment of advantageous mutations whose very advantage presupposes their existence.

At the genic level, therefore, the evolutionary adjustment by natural selection must be much more delicate and minute than it appears to be at the phenotypic level.

The picture of evolution that we have arrived at here is that in a mammalian species, new mutations due to base substitution occur in each member of the population at the rate of some half dozen per year, and in the case of a human population consisting of ten thousand breeding individuals, well over a million new molecular mutations will be added to the genetic pool in each generation. A majority of them are selectively neutral and their fate is largely controlled by random drift due to sampling of gametes in reproduction. They form a great flow of genetic variation that is undetectable at the ordinary phenotypic level. Thus, random genetic drift is at work perhaps on much grander scale than originally conceived by Wright (1931). In this state, each individual will be heterozygous in some four million nucleotide sites.

From time to time, for example, once every 300 generations or so in the case of horotelic (standard rate) evolution, a new environmental challenge forces the population to carry out gene substitution of definite adaptive nature, for which the population must pay "the cost of natural selection" (Haldane 1957). Evolutionary progress in the strict sense must have been achieved largely in this way.

Irrespective of such evolutionary episodes, it is expected that the great flow of genetic variation will cause all the lines to change their genic constitution at a relatively constant rate. Thus, after a lapse of time, each gene of "living fossils" will have undergone almost as many changes in terms of base replacement as the genes of their more progressive relatives. In this way, the footprints of time should be evident in all informational macromolecules.

CHAPTER THREE

# Effective Population Number

Random sampling of gametes in reproduction, as we have seen in the preceding chapters, has a very strong influence on the behavior of individual mutant genes in a population. In fact, even for a slightly advantageous mutant, its fate is largely determined by this factor.

More generally, in any finite population, gene frequencies are subject to stochastic change due to random sampling of gametes. Such a sampling process is a consequence of the fact that, in nature, a population cannot expand indefinitely, and, of the very large number of gametes produced each generation, only a tiny fraction can participate in forming the next generation.

For mathematical investigation of the cumulative effect of such sampling on the genetic constitution of a population, it is convenient to assume a simple population structure consisting of $N$ breeding individuals produced each generation by random union of $N$ male and $N$ female gametes extracted as random samples from the gene pool of the previous generation.

On the other hand, an actual population is likely to have a more complicated breeding structure, and it is desirable to reduce such a complicated situation to an equivalent simple case for which the mathematical treatment is much easier.

The concept of effective population number introduced by Wright (1931) in relation to evolutionary theory meets this need and has proved to be very useful in population genetics theory.

He has shown that the effective size may differ greatly from the apparent size, and usually it is much less. For example, if the population consists of $N_m$ males and $N_f$ females, the effective number is

$$N_e = \frac{4N_m N_f}{N_m + N_f}. \tag{1}$$

Unless the numbers of males and females are equal, this is always smaller than the actual number $N_m + N_f$. Especially, if the numbers of mature males and females differ greatly, $N_e$ depends mainly on the less numerous sex. In an extreme case, for example, if the herd is headed by one male, the effective size is roughly 4, even when the number of females is very large.

Wright (1938b) also derived the important result that if $N$ parents furnish varying numbers ($k$) of gametes to the next generation and if the population is stationary in size ($\bar{k} = 2$) and mating is at random, the effective number is

$$N_e = \frac{4N - 2}{V_k + 2}, \tag{2}$$

where $V_k$ is the variance of $k$. Thus, if the progeny number follows the Poisson distribution ($V_k = \bar{k} = 2$), the effective number becomes approximately equal to the actual number, i.e. $N = N_e$. If the variance is larger than the mean, as in most populations, the effective number becomes less. In an extreme situation, if all but one of the parents are sterile in a monoecious population, $V_k = (2N)^2/N - 2^2 = 4N - 4$ so that the effective number becomes only one ($N_e = 1$), as might be expected. On the other hand, if all the parents leave exactly the same number of offspring ($V_k = 0$), the effective number becomes about twice the actual number, i.e. $N_e = 2N - 1$. Such a situation is unlikely to occur in nature but may be realized in experimental populations under human control.

Probably one of the most important causes in nature which makes the effective number much smaller than the actual number is periodic reduction in the number of breeding individuals. Wright (1938b, 1939) showed that if the number of individuals changes cyclically with a relatively short period of $n$ generations, the effective number is

$$N_e = \tilde{N}, \tag{3}$$

where $\tilde{N} = n / \sum_{i=1}^{n} N_i$ is the harmonic mean (reciprocal of the mean of the reciprocals) of the number of individuals over one cycle. Thus the effective number is controlled largely by the phase of small numbers. For example, if an insect population expands in five generations from 10 to $10^6$ in geometric series and then returns to the initial size through the following five generations, the effective size is about 54.

An important improvement in the concept of effective population number has been brought about by Crow (1954), who examined the concept using three different but related properties of a finite population, that is, (i) the inbreeding effect, (ii) increase of variance in the distribution of gene frequencies and (iii) random extinction of alleles. Each of these may be used to define an effective population number. His idea was later extended and refined by Crow and Morton (1955) and also by Kimura and Crow (1963a). For a population which remains constant in size and is mating at random, these three effective numbers agree with each other, but in some extreme situations, they can be very different.

So, we shall examine briefly the properties of these effective numbers.

### (1) *Inbreeding Effective Number*

In a finite population, the inbreeding coefficient $f$, or the probability of a randomly chosen pair of homologous genes being descended from a common ancestral gene, increases with time. The inbreeding effective number denoted by $N_{e(f)}$ may be defined, in a monoecious population, as the reciprocal of the probability that two uniting gametes come from the same parent. In a population with separate sexes, two genes in the uniting gametes cannot come from the same parent but they might come from the same grandparent. So, in this case, we may define $N_{e(f)}$ as the reciprocal of such probability with respect to grandparents.

It can then be shown (cf. Kimura and Crow 1963a) that, for a monoecious population mating at random, the inbreeding effective number at generation $t$ is

$$N_{e(f)} = \frac{N_{t-1}\bar{k} - 1}{\bar{k} - 1 + V_k/\bar{k}}, \tag{4}$$

where $N_{t-1}$ is the number of individuals in the previous $(t-1)$th generation, each of which contributes a variable number, $k$, of gametes to the $t$th generation. This formula reduces to Wright's formula (2) if the population is stationary in size, i.e. $N_{t-1} = N$, $\bar{k} = 2$.

The corresponding formula for a random mating population with separate sexes is

$$N_{e(f)} = \frac{N_{t-2}\bar{k} - 2}{\bar{k} - 1 + V_k/\bar{k}}, \tag{5}$$

where $N_{t-2}$ is the number of individuals two generations back and $k$ refers to the number of gametes contributed by them. For a population of constant size, $N_{t-2} = N$, $\bar{k} = 2$ and we have

$$N_{e(f)} = \frac{4N - 4}{V_k + 2}, \tag{6}$$

which does not differ much from (2) if $N$ is large.

It can also be shown that in an ideal population consisting of $N_m$ males and $N_f$ females, the formula reduces to (1).

Actually, this simple and quite useful formula by Wright (formula 1) can be derived directly by using the definition of the inbreeding effective number as follows. The probability that two homologous genes in generation $t$ come from the same male in generation $t-2$ is $1/(4N_m)$, because the probability of both genes coming from males in generation $t-2$ is $1/4$, and the conditional probability of both coming from the same male, given that they both come from males, is $1/N_m$. Similarly, the probability that both come from the same female is $1/(4N_f)$. Thus the probability of both genes coming from the same grandparent is

$$\frac{1}{4N_m}+\frac{1}{4N_f}=\frac{N_m+N_f}{4N_mN_f},$$

which, when equated to $1/N_{e(f)}$, gives formula (1).

An important property of the inbreeding effective number is that it is related to the number of parents or grandparents rather than to that of the offspring.

(2) *Variance Effective Number*

In an ideal population consisting of $N$ breeding individuals, if the frequencies of a pair of alleles $A_1$ and $A_2$ are $p$ and $1-p$, the variance $V_{\delta p}$ of the change in gene frequency due to random sampling of gametes in one generation is $p(1-p)/(2N)$. It is natural, therefore, to define the variance effective number

$$N_{e(v)}=\frac{p(1-p)}{2V_{\delta p}}. \tag{7}$$

With this definition, it can be shown (cf. Kimura and Crow 1963a) that for a monoecious population, the variance effective number at the $t$th generation is

$$N_e=\frac{2N_t}{1-\alpha_{t-1}+(1+\alpha_{t-1})s_k{}^2/\bar{k}}, \tag{8}$$

where

$$s_k^2 = \frac{\Sigma(k - \bar{k})^2}{N_{t-1} - 1} = \frac{N_{t-1}V_k}{N_{t-1} - 1},$$

in which $k$ is the number of gametes contributed by $N_{t-1}$ parents (the subscript $t - 1$ refers to the $(t - 1)$th generation). In the above formula, $\alpha_{t-1}$ is a measure of the departure from Hardy-Weinberg proportions of genotypes, and is formally equivalent to the inbreeding coefficient. If the parent generation was derived from random mating and if $\alpha_{t-1}$ is not known, we substitute the expected value $-1/(2N_{t-1} - 1)$ for $\alpha_{t-1}$, giving

$$N_{e(v)} = \frac{(2N_{t-1} - 1)\bar{k}}{2(1 + V_k/\bar{k})}. \tag{9}$$

It can also be shown that for a population with separate sexes, if the $\alpha$'s and the progeny distributions are the same in both sexes, the corresponding formula is

$$N_{e(v)} = \frac{(N_{t-1} - 1)\bar{k}}{1 + V_k/\bar{k}}. \tag{10}$$

When the population size remains constant from generation to generation, formulae (9) and (10) reduce respectively to (2) and (6). Therefore the variance and inbreeding effective numbers agree with each other.

However, if the population size is not constant, these two effective numbers may be different, and in some extreme situations, they can be very different. For example, in a monoecious population decreasing in size, if each parent produces exactly one offspring ($\bar{k} = 1$, $V_k = 0$), we have $N_{e(f)} = \infty$ from (4) but $N_{e(v)} = N_{t-1} - 1/2 \approx 2N_t$ from (9). This big difference in the two effective numbers may easily be understood by noting that, in this example, no two individuals have a common ancestor and therefore the inbreeding effective number is infinity, while the sampling variance is not zero, and in fact, because of $V_k = 0$, the variance effective number becomes roughly twice the actual number of offspring. Another extreme example is a single plant producing by self-fertilization an infinite number of offspring so that $N_{t-1} = 1$, $V_k = 0$ and $\bar{k} = \infty$. Then $N_{e(f)} = 1$ from (4) but $N_{e(v)} = \infty$ from (9). This can again be understood readily by noting that in this example there is no sampling variance and therefore the variance effective number is infinity, while the inbreeding effect is equivalent to that of self-fertilization and therefore the inbreeding effective number is only one.

In general, the inbreeding effective number is related to the number of parents (or grandparents), while the variance effective number is related to that of offspring.

### (3) *Extinction Effective Number*

The ultimate rate by which one of the alleles is lost from the population by random drift is related to the size of the population and this may be used to define an effective size. Namely, by equating the rate of steady decay of population variability with the corresponding rate in an ideal population, i.e. $1/(2N)$, we can obtain the formula for the effective size. This was first done by Crow (1954), who, using a result by Haldane (1939), derived the formula

$$N_{e(R)} = \frac{4N}{V_k + 2}, \tag{11}$$

where $N_{e(R)}$ is the effective number with respect to random extinction. The formula is valid for a population of constant size. Essentially the same definition as the above has been used by Ewens (1969) to define the effective size.

Formula (11) is asymptotically equal to formulae (6) and (2), showing that for a population of constant size, the three effective numbers are essentially equivalent.

An interesting application of the extinction effective number has recently been made by Maruyama (1971b). As we have seen in Chapter 1, the average number of generations until fixation of a neutral mutant gene in a panmictic population is $4N_e$, where $N_e$ refers to the variance effective number. He has shown that in a subdivided population with circular arrangement of colonies and with migration taking place between adjacent colonies, the average number of generations until fixation is again given by $4N_e$, if we equate $1/(2N_e)$ with the rate of steady decay of the genetic variability of the entire population. Later in this book (see Chapter 8), we shall discuss some problems relating to the rate of decrease of heterozygosity in a subdivided population.

So far we have considered the problem of determining the effective population number when the mean and the variance of the progeny distribution have been measured. In contrast to such a retrospective approach, Robertson (1961) studied the problem of predicting the effect of artificial selection in reducing the population number. With selection, the expected number of offspring among different individuals must be different, and this makes the effective number smaller than the actual number.

Suppose that the expected number of offspring from a particular pair of parents is $K$, and that the values of $K$ among different pairs of parents are distributed with mean $\bar{K}$ and variance $V_K$. Then, Robertson showed that the effective number is given by

$$N_e = \frac{N + C^2}{1 + C^2}, \qquad (12)$$

where $C = \sqrt{V_K}/\bar{K}$ is the coefficient of variation of $K$. $C^2$ is approximately equal to the variance in fitness measured in logarithmic selective values. Note that if there is no selection, $C = 0$ and the formula reduces to $N_e = N$. Namely, in this formula, the population is regarded as "ideal" if there is no selection.

Nei and Murata (1966) combined Robertson's idea with formula (8) by Crow and Kimura to derive a new formula for the effective number when fertility is inherited. They obtained

$$N_e = \frac{N}{(1 + 3h^2)C^2 + 1/\bar{k}}, \tag{13}$$

where $h^2$ is the heritability of progeny number. If the population is stationary in size, $\bar{k} = 2$ and the above formula reduces to

$$N_e = \frac{4N}{(1 + 3h^2)V_k + 2}. \tag{14}$$

This is valid for a monoecious population, but it can also be applied to a population with separate sexes if the set of genes exerts the same effect on male and female fertilities. Taking heritability $h^2 = 0.3$ and variance of progeny number $V_k = 3$ (adjusted to $\bar{k} = 2$), both of which are estimates from data for female fertility in man, Nei and Murata obtained, from formula (14),

$$N_e \approx 0.52N.$$

Namely, the effective number is roughly half of the actual number.

All the above calculations are based on a model of discrete generation time. On the other hand, many organisms, including man, have a population structure in which individuals of all ages are present simultaneously and reproductive age varies. Therefore, it is desirable to investigate the problem of effective number for the case of overlapping generations. Kimura and Crow (1963a) studied this problem and derived an approximate formula for the case of continuous generation time, but their formula is unsatisfactory. In particular, it gives a very large effective number if the population contains a large number of non-reproducing individuals of higher ages.

Recently, Felsenstein (1969) carried out a rigorous analysis assuming a haploid population with overlapping generations and obtained the formula for the variance effective number,

$$N_e = \frac{BT}{1+K}, \tag{15}$$

where $B$ is the number of newborn individuals per year, $T$ is the mean age of mothers of newborns, and $K$ a quantity that may be roughly thought of as the ratio of death to birth rates during the reproductive period. His formula shows that, roughly speaking, the effective number does not differ greatly from the total number of newborn individuals during one generation.

Stimulated by his work, Crow and Kimura (1971) derived a new formula for the variance effective number that may be applied to a diploid population with continuous generations. It involves some approximations, but they believe that the new formula is sufficiently simple and realistic to be useful in analyzing human populations and that it can replace their earlier, incorrect formula.

In what follows we simply summarize their new result. Let $N_y$ dy be the number of individuals in the age interval $y$ to $y$ + dy, $\ell_y$ be the probability of surviving from birth to age $y$, and $b_y$ dy be the expected number of births to an individual in the age interval $y$ to $y$ + dy. Then

$$N_e = \overline{\ell} \int N_y v_y \, dy, \tag{16}$$

where

$$v_y = \frac{1}{e^{-my}\ell_y} \int^{\infty} e^{-mt} \ell_t b_t \, dt \tag{17}$$

is Fisher's reproductive value at age $y$ and

$$\overline{\ell} = \frac{\int_0^\infty \ell_y^2 b_y e^{-my}\,\mathrm{d}y}{\int_0^\infty \ell_y b_y e^{-my}\,\mathrm{d}y} \tag{18}$$

is the average probability in a cohort of surviving to age $y$, weighted by the proportion of total reproduction that occurs at that age and with each birth expressed as its present value. In these formulae, $m$ is the Malthusian parameter of population, defined by

$$\int_0^\infty e^{-mx} \ell_x b_x \,\mathrm{d}x = 1, \tag{19}$$

and integration is over all ages. If the census is taken periodically (say, yearly), summation may replace integration. An interesting point that Crow and Kimura noted is that although the formula (16) was derived through variance consideration it also represents the inbreeding effective number.

When the population is stationary in size, (16) reduces to

$$N_e = N_0 \tau \overline{\ell} \tag{20}$$

where $N_0$ is the number born each year and $\tau$ is the average age of reproduction, i.e.

$$\tau = \int_0^\infty t \ell_t b_t \,\mathrm{d}t.$$

This equation for the effective number is similar to the formula obtained earlier by Nei and Imaizumi (1966b), i.e.

$$N_e = N_m \tau, \tag{21}$$

where $N_m$ is the number born per year who survive until the mean age of reproduction and $\tau$ is the mean age of reproduction. When the birth rate is uniform and the death rate is low during the period of reproduction, formulae (20) and (21) become essentially the same.

CHAPTER FOUR

# Natural Selection and Genetic Loads

The term genetic load owes its origin and popularity to H. J. Muller's article "Our load of mutations" (Muller 1950). With the great insight and originality so characteristic of him, Muller amassed diverse evidence to lead to the conclusion that gene mutation is an important ultimate cause of impairment of our health (contrary to the opinion of many medical men at that time) and that each of us carries on the average some eight detrimental genes in heterozygous condition which express their deleterious effect due to a small amount of dominance.

That his conjecture was essentially correct was later confirmed by the work of Morton, Crow and Muller (1956) through the study of inbreeding depression in man. They showed that a typical individual carries lethal and detrimental genes amounting to 1.5 ~ 2.5 "lethal equivalent" per gamete. A lethal equivalent is a group of genes of such number that, if dispersed in different randomly chosen individuals, they would cause on the average one death (cf. Crow and Kimura 1963). Thus one lethal equivalent may consist of one lethal gene, two semilethal genes with 50% viability, etc. On the other hand, an extensive study of inbreeding by Neel and Schull (1962) in Hiroshima and Nagasaki gave a lower and less consistent result (cf. Schull and Neel 1965), although still more recent and careful study in Kyushu by Yanase's group (cf. Yamaguchi et al., 1970) gave 0.62 ~ 0.69 lethal equivalents per gamete. It is not surprising that contemporary values are smaller than those given by Morton, Crow and Muller, which were based on data from previous generations. With improved environment, better health care, and higher living standards the probability that a mutant will cause death is lessened and this is reflected in a smaller number of lethal equivalents revealed by inbreeding.

In *Drosophila melanogaster*, according to Greenberg and Crow (1960), the typical individual carries two lethal equivalents. It may perhaps be surprising to realize that each of us, even if he appears to be healthy, carries in heterozygous condition a number of detrimental genes that is enough to kill each of us at least once if suddenly made homozygous.

Muller (1950) also derived a very interesting principle that a slightly detrimental mutant gene causes in the long run as much biological damage to the population as a highly detrimental mutant. In other words, each detrimental mutant gene eventually produces one genetic death.

In his article, Muller used the term load in the sense of burden or handicap due to detrimental genes, as is still used by some geneticists. However, in this book, we shall use the term genetic load in a more restricted sense, which is due to Crow (1958). He proposed this term as a measure of genotypic selection intensity. According to Crow (1958), the genetic load is defined as the proportion by which the fitness of the average genotype in the population is reduced in comparison to the best genotype. If the fitness is expressed in terms of Wright's selective values ($w$), the genetic load is

$$L = \frac{w_{max} - \bar{w}}{w_{max}} \tag{1}$$

where $\bar{w}$ is the average selective value of the population and $w_{max}$ is that of the optimum genotype. For the model of discrete generation time, $w$ is the appropriate measure of fitness since it represents the number of offspring that the average individual contributes to the next generation.

For the model of continuous generation time in which the fitness is measured in Malthusian parameters $m$ (Fisher 1930b), that is to say, the geometric rate of population growth, the load is

$$L = m_{\max} - \bar{m} \tag{2}$$

where $m_{\max}$ is the Malthusian parameter of the optimum genotype and $\bar{m}$ is that of the population average (Kimura 1960a). The genetic load thus defined represents the amount of selective elimination due to genotypic differences and therefore is adequate to measure the genotypic selection intensity.

We owe the mathematical formulation underlying the concept of genetic load chiefly to Haldane (1937, 1957), though a number of studies have since been done to investigate various cases such as finite population size (Kimura, Maruyama and Crow 1963, Kimura and Crow 1964), epistatic gene interaction in fitness (Kimura and Maruyama 1966), mother-child incompatibility (Crow and Morton 1960), segregation distortion (Kimura 1960b, Kimura and Kayano 1961, Crow and Kimura 1970), intermediate optimum in quantitative characters (Kimura 1965b), deviation from equilibrium due to random drift (Robertson 1970a, Kimura and Ohta 1970b) and so on. Recently, an excellent review on the theoretical aspects of the genetic loads was presented by Crow (1970). Also, ecological consideration of the problem will lead to new insights into the action of natural selection in biological systems.

In the following sections, we shall discuss several kinds of genetic loads.

## THE MUTATIONAL LOAD

The mutational load is the load created by the elimination of recurrent harmful mutations.

It was shown first by Haldane (1937) that the decrease of fitness due to recessive mutation is equal to the mutation rate $u$. For semidominant mutations, the decrease of fitness (i.e. the load) is $2u$. An important point to note is that the load is almost independent of the degree of the harmfulness of mutants. The same conclusion was obtained by Muller (1950) in terms of "genetic death." A simple way of seeing this is to consider a particular locus in a haploid population in which a mutant gene $A_2$ is produced at the rate $u$ per generation from the normal gene $A_1$. We shall first use a model of discrete generation time and denote the relative fitnesses of $A_1$ and $A_2$ by 1 and $1 - s$ respectively, where $s$ is the selection coefficient against the mutant. Throughout this book we shall use the symbol $w$ to denote the selective value in the sense of Wright (cf. Wright 1969), namely, the reproductive value of a gene or genotype in the discrete generation model. If $p$ is the frequency of the mutant $A_2$, the amount of change by natural selection in $p$ per generation is

$$\Delta p = \frac{(1-s)p}{\bar{w}} - p = -\frac{sp(1-p)}{\bar{w}},$$

where $\bar{w} = (1-p) + p(1-s) = 1 - sp$ is the mean fitness of the population (Table 4.1).

TABLE 4.1. Fitnesses in selective values and Malthusian parameters for a haploid population.

| Genotype | $A_1$ | $A_2$ |
|---|---|---|
| Frequency | $1-p$ | $p$ |
| Relative fitness (selective value) discrete generation | 1 | $1-s$ |
| Relative fitness (Malthusian parameter) continuous generation | 0 | $-s$ |

At equilibrium, this decrease is balanced by the mutational production of $A_2$ from $A_1$ so that

$$\Delta p = -\frac{sp(1-p)}{1-sp} + u(1-p) = 0. \tag{3}$$

We neglect the reverse mutation rate, as this gives only an insignificant effect on $\Delta p$ when $p$ is small. Then, at equilibrium,

$$\hat{p} = \frac{u}{(1+u)s}.$$

Since $w_{\max} = 1$, $\bar{w} = 1 - sp$ in the present case, from the definition, the mutational load at equilibrium is

$$L_{\text{mut}} = \frac{1-(1-s\hat{p})}{1} = s\hat{p} = \frac{u}{1+u} \approx u, \tag{4}$$

since $u$ is much smaller than unity. That is to say, the mutational load is equal to the mutation rate per gamete.

In the above treatment, we have considered relative frequencies and relative fitnesses. A question that naturally arises is whether the above conclusion on mutational load will not be altered if the population regulating mechanism is taken into account. Certainly, the absolute fitness of a gene in terms of the number of offspring per individual must depend on population density. To investigate this problem, it may be convenient to use a continuous model of population growth and measure fitness in Malthusian parameters, i.e. in terms of geometric growth rates. Let $n_1$ and $n_2$ be the absolute numbers of $A_1$ and $A_2$ in a haploid population and let $N = n_1 + n_2$ be the total population number. We shall denote by $m_1$ and $m_2$ the Malthusian parameters of $A_1$ and $A_2$, i.e.

$$\frac{1}{n_1}\frac{\mathrm{d}n_1}{\mathrm{d}t} = m_1 \qquad \text{and} \qquad \frac{1}{n_2}\frac{\mathrm{d}n_2}{\mathrm{d}t} = m_2.$$

Generally, both $m_1$ and $m_2$ may depend on $n_1$ and $n_2$. For example, under a logistic population growth, if we assume that the mutant $A_2$ has lower intrinsic reproductive rate but is equally resistant to overcrowding, $m_1$ and $m_2$ may be expressed as follows;

$$m_1 = \alpha_1 - c\bar{\alpha}N \qquad \text{and} \qquad m_2 = \alpha_2 - c\bar{\alpha}N,$$

in which $\alpha_1$ and $\alpha_2$ are intrinsic fitnesses of $A_1$ and $A_2$ such that $\alpha_1 > \alpha_2$ and $\bar{\alpha} = (n_1\alpha_1 + n_2\alpha_2)/N$, the mean of two $\alpha$'s (cf. Kimura and Crow 1969). The actual Malthusian parameter of $A_2$ may be positive when $N$ is small but becomes negative when $N$ is large. The total population number is regulated by the equation

$$\frac{dN}{dt} = N(\bar{\alpha} - cN),$$

while the rate of change of the logarithmic gene ratio by selection is

$$\frac{d}{dt} \log_e \left(\frac{p_1}{p_2}\right) = \alpha_1 - \alpha_2.$$

In this case, we may call $s = \alpha_1 - \alpha_2$ the selective advantage of $A_1$ over $A_2$ measured in Malthusian parameters. In actual populations, regulation of the population number may be more intricate. So, in the following treatment, we shall only assume that the mutant gene is less fit than its normal allele ($m_1 > m_2$).

Under recurrent mutation from $A_1$ to $A_2$, we have

$$\frac{dn_1}{dt} = m_1 n_1 - u n_1$$

and

$$\frac{dn_2}{dt} = m_2 n_2 + u n_1,$$

where $u$ is the mutation rate from $A_1$ to $A_2$. Thus at equilibrium, in which the production of deleterious genes by mutation is counterbalanced by selective elimination, we have

$$\hat{m}_1 = u$$

$$\hat{m}_2 = \frac{-u\hat{n}_1}{\hat{n}_2} \tag{5}$$

where the symbol ˆ on each letter indicates its value at equilibrium. Using the definition given as formula (2), the mutational load at equilibrium becomes

$$L_{\text{mut}} = \hat{m}_1 - \hat{\bar{m}} = \hat{m}_1 - \frac{\hat{n}_1\hat{m}_1 + \hat{n}_2\hat{m}_2}{\hat{n}_1 + \hat{n}_2} = \frac{(\hat{m}_1 - \hat{m}_2)\hat{n}_2}{\hat{n}_1 + \hat{n}_2}$$

which reduces to

$$L_{\text{mut}} = u \tag{6}$$

if we apply equations (5). This shows that as long as the mutation is unconditionally deleterious, the population regulating mechanism has no effect on the mutational load.

For a diploid population, the treatment is more complicated, but it was shown by Kimura (1961b) that if fitnesses (in terms of selective values) of the three genotypes $A_1A_1$, $A_1A_2$ and $A_2A_2$ are respectively 1, $1 - hs$ and $1 - s$, then in a random mating population, the mutational load is given approximately by

$$L_{\text{mut}} = u\{1 - \theta \pm \sqrt{\theta(2 + \theta)}\}, \tag{7}$$

where

$$\theta = \frac{sh^2}{2u(1 - 2h)}. \tag{8}$$

In the above formula for $L_{\text{mut}}$, the plus sign is taken when $h$ (dominance) $< 0.5$ and the minus sign when $h > 0.5$. Figure 4.1 shows the value of $L_{\text{mut}}$ for lethal mutations ($s = 1$) as a function of $h$ (degree of dominance) assuming mutation rate $u = 10^{-5}$.

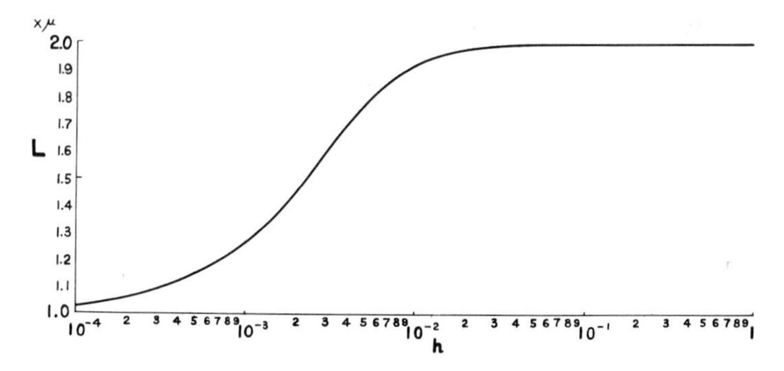


FIGURE 4.1. The load ($L$) due to lethal mutations as a function of the degree of dominance. Ordinate: mutational load $L$ expressed with mutation rate $\mu$ ($= 10^{-5}$) as the unit; abscissa: degree of dominance $h$. (From Kimura 1961.)

From the figure it may be seen that the load increases continuously from $u$ to approximately $2u$ as $h$ changes from 0 (complete recessivity) to 1 (complete dominance). This agrees with the intuitive argument that if the mutant gene is dominant each mutation leads to one genetic death, while if it is recessive two mutations are required to lead to one genetic death. We may note that $L_{\text{mut}}$ theoretically exceeds $2u$ when $h > 0.5$. However, the amount of excess is usually so small that it may be neglected. In general, if $|\theta|$ in (8) is large, we have

$$L_{\text{mut}} = u\left(2 - \frac{1}{2\theta}\right). \tag{9}$$

Thus, if $|\theta|$ is very large or $sh^2$ is much larger than $2u \times (|1 - 2h|)$, we have approximately

$$L_{\text{mut}} = 2u, \tag{10}$$

For lethal and semilethal genes, mutation rates and degree of dominance are roughly $u = 10^{-5}$ and $h \approx 0.02$ (cf. Crow and Temin 1964), so that $\theta$ is probably 10 or more. Therefore (10) serves as a simple but a realistic approximation. Formula (10) means that the mutational load is equal to the mutation rate per zygote. Crow (1957) called this rule the Haldane-Muller principle, and it has been used widely as a basis for assessing the genetic damage of mutagenic agents, such as ionizing radiation, to human populations.

The Haldane-Muller principle should have a wide applicability since in man a majority of visible mutations seem to have an appreciable effect in heterozygous condition (cf. Muller 1950, Stevenson 1958).

However, there are at least two circumstances under which this principle needs some modification. One is the case in which there is epistatic gene interaction in fitness and another is the random gene frequency drift in a small population. First, consider all the loci capable of producing deleterious mutations. If there is no epistatic interaction in fitness, the total mutational load is the sum of the loads at individual loci. However, when we consider their physiological effect on an individual, it is reasonable to assume that there is "reinforcing type" epistasis between loci. Namely, the detrimental effect of two or more mutant genes collectively is larger than the sum of the detrimental effects of individual mutants. Kimura and Maruyama (1966) investigated a model in which the deleterious effect of mutant genes to an individual is given by the quadratic expression of the number of mutant genes. Such a model may be realistic if the phenotypic suppression of mutational damage by developmental homeostasis breaks down rapidly as the number of mutant genes increases. In particular it was shown that if the deleterious effect is proportional to the square of the number of mutant genes, the load under random mating becomes roughly half as large as in the case of no epistasis, provided that the average number of such genes per individual is fairly large. Recent work of Mukai (1964, 1969a) has revealed that in *Drosophila* the mutation rate for slightly deleterious genes, which he called "viability polygenes" (and which have some 2% disadvantage in the homozygote), is much higher than previously considered. The total mutation rate for such genes amounts to at least 70% per individual. Mukai has also obtained evidence for suggesting the existence of quadratic interaction in fitness, so the total mutation load due to such genes is probably lower than 0.3.

The above calculations are all based on the assumption that random elements play no significant role in determining gene frequencies. However, in a small population, it is expected that the load would increase because the gene frequencies tend to drift away from equilibrium values. This problem was investigated by Kimura, Maruyama and Crow (1963) and it was shown that in small populations the load is considerably larger than in a large population. An interesting result which emerged from the study is that, for a wide range of population sizes, a mutant that is slightly harmful is more damaging to the fitness of the population than a mutant with much greater harmful effect.

## SEGREGATIONAL LOAD

The term segregational load was used by Crow (1958) to denote the load due to heterozygote superiority in fitness (overdominance). The load is created because in each generation inferior homozygotes are produced by segregation. In this book we shall use this term in a wider sense (cf. Kimura 1960b) to include also the load due to segregation distortion.

### A. *The Overdominance Load*

If two alleles, $A_1$ and $A_2$, are overdominant, both alleles may be maintained in a population in stable equilibrium. This may be seen by considering an extreme situation in which one of the alleles happens to be rare. Then, this rare allele occurs almost exclusively in heterozygotes under random mating, and therefore it is on the average more advantageous than the other allele which, because of higher frequency, occurs not only in heterozygotes, but also a considerable fraction of the time in less fit homozygotes. In other words, each allele tends to increase when rare, thus being prevented from loss.

A rigorous mathematical demonstration that overdominant alleles will be maintained in a population in stable equilibrium was first given by Fisher (1922), assuming a pair of alleles.

Since then, overdominance has become the most popular mechanism assumed to explain balanced polymorphism.

Let $p_1$ and $p_2$ $(= 1 - p_1)$ be respectively the frequencies of $A_1$ and $A_2$ in the population, and let $s_1$ and $s_2$ be the selection coefficients against homozygotes, $A_1A_1$ and $A_2A_2$, such that if $w_{12}$ is the fitness (measured in Wright's selective values) of the heterozygote ($A_1A_2$) then those of $A_1A_1$ and $A_2A_2$ are respectively $w_{12}(1 - s_1)$ and $w_{12}(1 - s_2)$. We assume that the selection coefficients are positive and constant. Under random mating (Table 4.2), the equilibrium is reached when $s_1p_1 = s_2p_2$ so that

$$\hat{p}_1 = \frac{s_2}{s_1 + s_2} \tag{11}$$

is the equilibrium frequency of $A_1$.

If we denote by $\bar{w}$ the mean fitness of individuals in the population, i.e.

$$\begin{aligned} \bar{w} &= w_{12}\{(1 - s_1)p_1^2 + 2p_1p_2 + (1 - s_2)p_2^2\} \\ &= w_{12}\left\{\left(1 - \frac{s_1s_2}{s_1 + s_2}\right) - (s_1 + s_2)(p_1 - \hat{p}_1)^2\right\}, \end{aligned} \tag{12}$$

then it may easily be seen that $\bar{w}$, as a function of $p_1$, takes a relative maximum value at the equilibrium point. At this

TABLE 4.2. Fitnesses and frequencies of genotypes with overdominance.

| Genotype | $A_1A_1$ | $A_1A_2$ | $A_2A_2$ |
|---|---|---|---|
| Fitness | $w_{12}(1 - s_1)$ | $w_{12}$ | $w_{12}(1 - s_2)$ |
| Frequency | $p_1^2$ | $2p_1p_2$ | $p_2^2$ |

equilibrium, the mean fitness is

$$\hat{\bar{w}} = w_{12}\left(1 - \frac{s_1 s_2}{s_1 + s_2}\right). \tag{13}$$

By comparing this with the fitness of the most fit genotype, i.e. with $w_{12}$, and using definition (1), we obtain the formula for the overdominance load in an infinite population:

$$L_{OD} = \frac{w_{12} - \hat{\bar{w}}}{w_{12}} = \frac{s_1 s_2}{s_1 + s_2}. \tag{14}$$

As pointed out by Crow (1958), this is equal to $s_1\hat{p}_1^2 + s_2\hat{p}_2^2$, i.e. the sum of the fractions of selective elimination of $A_1A_1$ and $A_2A_2$. For example, if $s_1 = 0.01$ and $s_2 = 0.99$, then $L_{OD} = 0.0099$, of which $s_1\hat{p}_1^2 = 0.01 \times (0.99)^2 = 0.009801$ and $s_2\hat{p}_2^2 = 0.99 \times (0.01)^2 = 0.000099$. In other words, the slightly deleterious allele $A_1$ (99% viability when homozygous) contributes some 100 times as much to the load as the nearly lethal allele $A_2$ (1% viability when homozygous).

Another interesting property that may be useful to assess the overdominance load in actual populations is that it is equal to the product of the selection coefficient against either one of the homozygotes and the frequency of the corresponding allele (Morton 1960). That is,

$$L_{OD} = s_1\hat{p}_1 = s_2\hat{p}_2.$$

With multiple alleles, $A_1, A_2, \ldots, A_n$, the problem becomes more complicated, but if all the heterozygotes have the same fitness and if $s_i$ is the selection coefficient against homozygote $A_iA_i$, then, as shown by Wright (1949), the frequency of the $i$th allele at equilibrium is

$$\hat{p}_i = \frac{1/s_i}{\sum_1^n (1/s_i)} \qquad (s_i > 0), \tag{15}$$

and the equilibrium is stable. Then, the overdominance load is

$$L_{OD} = \sum_1^n s_i \hat{p}_i^2 = \frac{1}{\sum_1^n (1/s_i)}. \tag{16}$$

Note that the load in this case is equal to $\tilde{s}/n$ where $\tilde{s}$ is the harmonic mean of the $n$ selection coefficients. This means that the load becomes smaller as the number of mutually heterotic alleles in a population increases. Still, the load may be estimated from the equilibrium frequency and the homozygous disadvantage of only one allele, since $L_{OD} = s_i \hat{p}_i$ is valid for any $i$ ($i = 1, 2, \ldots, n$). If the fitnesses of all the heterozygotes are not equal, this method gives a minimum estimate.

Generally, the overdominance load is expected to be much larger than the mutation load since the former is of the order of selection coefficients while the latter is of the order of mutation rates.

Under random mating and constant fitnesses of genotypes, the necessary and sufficient conditions for the maintenance of stable polymorphism by selection are quite simple if the number of segregating alleles is only two. Namely, the heterozygote must have higher fitness than either homozygote. With multiple alleles, however, the conditions become more complex.

For example, it is not necessary that all the heterozygotes be higher in fitness than any homozygote for the equilibrium to be stable. The required conditions, necessary and sufficient, may most readily be derived (Kimura 1956b) by using the principle that, under random mating and constant fitnesses of genotypes, the stable equilibrium corresponds to the relative maximum of the mean fitness of the population. Note that this principle does not hold for frequency-dependent selection.

Since overdominance is often assumed to be the main cause of polymorphism over a large number of loci in a genome, it may be of considerable importance to consider what the total load will be when two or more overdominant loci are segregating in a population.

Suppose that $n$ overdominant loci are segregating independently and that their contribution to the selective value is multiplicative such that on a logarithmic scale the fitness is additive between loci. Then, if $\ell_i$ is the load at the $i$th locus ($i = 1, 2, \ldots, n$), the average fitness of the population is

$$\prod_{i=1}^{n} (1 - \ell_i) \approx e^{-\Sigma \ell_i} \tag{17}$$

as compared with the individual which is heterozygous for all $n$ loci. Thus the total load in terms of selective values is

$$L_T = 1 - e^{-\Sigma \ell_i}. \tag{18}$$

In terms of Malthusian parameters (or in log $w$, approximately), the total load is $\Sigma \ell_i$, and it is often more convenient to express the load in this way. Note that when $\ell_i$ is small, the load in terms of log $w$ is approximately $\ell_i$, since $\log_e (1 - \ell_i) \approx -\ell_i$.

For example, if 100 overdominant loci are segregating, each producing a load of 0.01, then $\Sigma \ell_i = 100 \times 0.01 = 1$ so that $L_T = 1 - e^{-1} \approx 0.63$. If the number of overdominant loci is 10 times as large, $\Sigma \ell_i = 10$ and $1 - e^{-10} \approx 0.999954$. This is a very large load, and if the selection is carried out by premature death of less fit genotypes, each individual must produce on the average $e^{10}$ young (not considering deaths due to environmental causes), in order to keep the adult population number constant from generation to generation. This means that an enormous reproductive excess is needed to hold such a large number of strongly overdominant loci simultaneously segregating in the population. If the species cannot afford such a reproductive excess, the type of selection we have postulated cannot be exerted at each locus.

One way of avoiding such a heavy load is to assume that selection is based on competition between genotypes actually present in a population at a given moment. Under this type of selection, the fitness of the theoretically optimum individual which is heterozygous for all $n$ loci is irrelevant when $n$ is large, since such an individual is unlikely to appear in a finite population. For example, if $n = 40$, the expected frequency of such an all-heterozygous individual is less than $2^{-40}$ or one in a trillion. In such a case, we can calculate, for a given population number $N$, the fitness of the individual having the most probable largest number of heterozygous loci.

Let us consider a particular locus and let $s_1$ and $s_2$ be the selection coefficient against the two homozygotes, such that in logarithmic scale, the relative fitnesses of $A_1A_1$, $A_1A_2$ and $A_2A_2$ are $-s_1$, 0, and $-s_2$. Then, the contribution of this locus to the mean in fitness is

$$m = -s_1\hat{p}_1{}^2 - s_2\hat{p}_2{}^2 = -\frac{s_1s_2}{s_1 + s_2},$$

and to the variance in fitness is

$$\sigma^2 = s_1{}^2\hat{p}_1{}^2 + s_2{}^2\hat{p}_2{}^2 - m^2 = \left(\frac{s_1s_2}{s_1 + s_2}\right)^2.$$

Under multiplicative fitness between loci, the total fitness in log $w$ scale is distributed approximately normally with mean

$$M_T = \Sigma m = -\sum \frac{s_1 s_2}{s_1 + s_2} \tag{19}$$

and variance

$$\sigma_T^2 = \Sigma\sigma^2 = \sum \left(\frac{s_1 s_2}{s_1 + s_2}\right)^2, \tag{20}$$

where the summation is over all the overdominant loci. We may assume that the mean absolute fitness of the population on the logarithmic scale is always adjusted to 0 so that only deviation from $M_T$ matters. Using Frank's formula for "the most probable largest normal value" (cf. Gumbel 1958), we estimate that an individual having the most probable highest value in logarithmic fitness exceeds the mean by about

$$\tilde{x}_{N,1} \approx \sqrt{2 \log_e (0.4N)} \tag{21}$$

times of $\sigma_T$. Then, the load due to overdominance with respect to competition is

$$\tilde{L} = \sigma_T \tilde{x}_{N,1} = \sqrt{\sum \left(\frac{s_1 s_2}{s_1 + s_2}\right)^2 \times 2 \log_e (0.4N)}.$$

In terms of this load, the individual having the highest fitness in the population must produce $e^{\tilde{L}}$ times as many offspring as the population average (excluding environmental effects). For small values of $s_1$ and $s_2$, this requirement is not very stringent. For example, if $s_1 = s_2 = 0.01$, $n = 1{,}000$, $N = 25{,}000$, we have $\tilde{L} = \sqrt{0.46} \approx 0.68$ so that $e^{\tilde{L}} = e^{0.68} \approx 1.97$. However, this load increases rapidly as the selection coefficients become larger. For example, if the selection coefficients are 10 times as large ($s_1 = s_2 = 0.1$) but with equal numbers of loci and individuals, we have $\tilde{L} = \sqrt{46} = 6.8$. This means that the individual having the highest fitness in the population is likely to produce $e^{6.8} \approx 898$ times as many offspring as the population mean (again, not considering environmental deaths). However, this is unlikely for most mammalian species.

In order to avoid a heavy overdominance load as well as strong inbreeding depression in fitness when a large number of overdominant loci are segregating, Sved, Reed and Bodmer (1967) assumed that fitness rapidly reaches a modest upper limit as the number of heterozygous loci increases, and similarly, a modest lower limit as the number decreases. One shortcoming of their model is that the rate of inbreeding depression decreases as the inbreeding coefficient increases, contrary to most observational results.

A slightly different model was proposed by King (1967), who assumed that the underlying scale of fitness is a linear function of the number of heterozygous loci, and those individuals having a fitness score larger than a certain value survive to leave offspring. Namely, truncation selection is practiced in nature as in animal breeding based on the number of heterozygous loci of each individual. It can be shown that under this scheme of selection, a modest selective elimination such as 50% can still hold a large number of heterotic loci each with selection coefficient of 0.01. Whether such a truncation type model is realistic remains to be seen.

Linkage disequilibrium may reduce the overdominance load if it is sufficiently strong that practically only two complementary chromosome types are segregating within a single population. In this case, all the loci are organized, so to speak, into a super-gene. The situation is then analogous to a single pair of overdominant alleles segregating within a locus. Recently, Lewontin and his colleague (Franklin and Lewontin 1970) showed by computer simulation that under a multiplicative fitness model with overdominance at each locus a marked linkage disequilibrium will be realized if the recombination fraction between adjacent loci is restricted to a certain low range. If this effect turns out to be strong enough and the underlying condition is sufficiently realistic, it will be important in understanding the formation of supergenes.

We would like to point out here that if a very large number of overdominant loci (of the order of total population number) are segregating independently, a significant random fluctuation will be introduced into the selection coefficient at each locus. Let us suppose that there are $n$ overdominant loci at each of which a pair of overdominant alleles with symmetric effect is segregating (with selection coefficient $s$ against either homozygote). Take a particular locus and let $A_1$ and $A_2$ be a pair of alleles segregating in this locus. If we consider an individual with a particular combination of alleles, say $A_1A_1$ at this locus, there are $n - 1$ overdominant loci in its genetic background, and at each of these the relative selective value is either 1 or $1 - s$ with equal probability. In terms of logarithmic selective values, then, the genetic background has approximately a mean of $-(n - 1)(s/2)$ and a variance of $(n - 1)(s^2/4)$. If the frequency of $A_1$ happens to be $p$, then, there are $Np^2$ individuals of genotype $A_1A_1$, and therefore, the average fitness of $A_1A_1$ individual, in log $w$ scale, is distributed approximately normally with mean $-s - (s/2)(n - 1)$ and variance $(n - 1)s^2/(4Np^2)$. Similarly the average fitness of $A_1A_2$ individuals has mean $-(s/2)(n - 1)$ and variance $(n - 1)s^2/\{8Np(1 - p)\}$. Thus, if we denote by $s_1'$ the "apparent" selection coefficient against $A_1A_1$ at this locus, it is distributed approximately normally with the mean and variance given by

$$\overline{s_1'} = s$$

and

$$\sigma^2(s_1') = \frac{(n - 1)s^2(2 - p)}{8Np^2(1 - p)}. \qquad (22)$$

Therefore, the standard deviation

$$\sigma(s_1') = (s/p) \times \sqrt{(n-1)(2-p)/8N(1-p)}$$

may be much larger than the mean $s$, if $p$ is small or if $n$ is much larger than $N$. Similarly, the apparent selection coefficient, $s_2$, against $A_2A_2$ at this locus has mean $s$ and variance

$$\sigma^2(s_2') = \frac{(n-1)s^2(1+p)}{8N(1-p)^2 p}. \tag{23}$$

Such random fluctuation in the apparent fitness in each locus will make the estimation of intrinsic fitness very difficult and the evolutionary process quite capricious.

As noted earlier, the overdominance load becomes smaller as the number of alleles per locus increases. However, in a finite population, this number is limited by random drift caused by random sampling of gametes in reproduction, even if an infinite sequence of mutually heterotic alleles may be produced by mutation. This problem was investigated by Kimura and Crow (1964). Their results show that overdominance is quite inefficient as a factor for keeping a large number of alleles per locus in a finite population: In order to maintain a large number of alleles within an isolated population, a high mutation rate is required. An interesting application of the theory of Kimura and Crow was recently made by Kerr (1967) who investigated the number of sex-determining alleles in bee populations in Brazil.

### B. *The Distortional Load (the Load Due to Meiotic Drive)*

A gene that is unconditionally deleterious in zygotes may nevertheless be maintained in a population if it causes excessive segregation at meiosis or if gametes containing it have a selective advantage in fertilization.

Typical examples are $t$ alleles found in mouse and the SD factor in *Drosophila*.

Consider a pair of alleles, $a$ and $A$, and assume that $a$ is a recessive lethal but causes distorted segregation in heterozygous ($aA$) males.

We shall designate by $k$ ($k > 0.5$) the segregation proportion of allele $a$ among gametes produced by the $aA$ male; namely, the segregation ratio of $a$ to $A$ is $k:1-k$, instead of the ordinary ratio of $0.5:0.5$. Let $p^*$ and $p^{**}$ be the frequencies of $a$ in male and female gametic outputs. Under random mating, zygotic frequencies immediately after fertilization are

$$p^*p^{**}aa,\ [p^*(1-p^{**})+p^{**}(1-p^*)]aA,\qquad (1-p^*)(1-p^{**})AA. \quad (24)$$

Note that the Hardy-Weinberg law cannot be applied here because gene frequencies in male and female gametic outputs are different. The frequency of $a$ in the male gametic output for the next generation may be obtained by noting that $aa$ is lethal and the segregation ratio from $aA$ male is $k\ a:(1-k)\ A$.

$$p^{*\prime} = \frac{k(p^*+p^{**}-2p^*p^{**})}{1-p^*p^{**}}. \quad (25)$$

The corresponding frequency in the female gametic output may be obtained from the right side of the above equation by putting $k = 1/2$. Thus

$$p^{**\prime} : p^{*\prime} = 1/2 : k$$

or

$$p^{**\prime} = \frac{p^{*\prime}}{2k}. \quad (26)$$

At equilibrium, in which gene frequencies do not change

with generations (i.e. $p^{*\prime} = p^*$ and $p^{**\prime} = p^{**}$), we may drop the primes in the above formulae. This leads to

$$k\left(1 + \frac{1}{2k} - \frac{p^*}{K}\right) \Big/ \left(1 - \frac{p^{*2}}{2k}\right) = 1$$

or

$$p^{*2} - 2kp^* + k(2k - 1) = 0. \qquad (27)$$

The above quadratic equation for $p^*$ has two roots,

$$k + \sqrt{k(1-k)} \qquad \text{and} \qquad k - \sqrt{k(1-k)},$$

but, of these two, the latter is the pertinent one, reducing to 0 when $k = 1/2$ (i.e. in the case of no excessive segregation). Thus, the equilibrium frequency of $a$ in male gametic output is

$$\hat{p}^* = k - \sqrt{k(1-k)}. \qquad (28)$$

The corresponding value for the female gametic outputs is

$$\hat{p}^{**} = \frac{1}{2k}\,(k - \sqrt{k(1-k)}). \qquad (29)$$

The proportion of lethal individuals at equilibrium

$$\hat{p}^*\hat{p}^{**} = \frac{1}{2k}\,(k - \sqrt{k(1-k)})^2 = \frac{1}{2}\,(1 - 2\sqrt{k(1-k)})$$

gives the distortional load or the load due to segregation distortion (Kimura 1960b):

$$L_{SD} = 1/2(1 - 2\sqrt{k(1-k)}). \qquad (30)$$

The load $L_{SD}$ represents the intensity of genotypic selection that is acting in a population maintaining the lethal gene at a high frequency due to segregation distortion.

According to Dunn (1957), the segregation percentage of $t$ ranges from 89% to 99% ($k = 0.89$ to $0.99$), so $L_{SD}$ lies between 0.19 and 0.40. This means that in a wild population of mice, the load due to maintaining $t$ alleles amounts to about 20–40%. In a more general treatment in which the fitness of heterozygote *Aa* is assumed to be $1-h$ rather than 1, it has been shown (Crow 1970, Crow and Kimura 1970, p. 311) that the load is decreased when the lethal allele ($a$) is partially dominant ($0 < h < 1/2$).

A similar example of segregation distortion was found for a supernumerary chromosome called $f_\ell$ in *Lilium callosum,* a species of lily indigenous to Japan. The chromosome $f_\ell$ has a tendency to increase its number due to preferential segregation in embryo-sac mother cells, but has the deleterious effect of reducing both pollen and seed fertility when more than one is present. In wild populations of this lily in Kyushu, the $f_\ell$ chromosome is found in more than 70% of individuals in a population, on the average. It was shown that the distortional load for maintaining the supernumerary chromosome amounts to about 17% (Kimura and Kayano 1961).

## DRIFT AND DYSMETRIC LOADS

As pointed out in a preceding section, if overdominant alleles are segregating at a locus, the mean fitness $\bar{w}$ as a function of allelic frequencies takes a relative maximum at the stable equilibrium point. This holds for a random mating population of an infinite size, but in a finite population, the allelic frequencies tend to deviate from this equilibrium point due to random frequency drift, and the mean fitness is thereby reduced. We shall call such a load the drift load. In symbols,

$$L_{\text{drift}} = \frac{\hat{\bar{w}} - E(\bar{w})}{\hat{\bar{w}}}, \tag{31}$$

where $\hat{\bar{w}}$ is the mean fitness of individuals in the population when allelic frequencies correspond to the equilibrium values (in an infinite population) and $E(\bar{w})$ is the expected value of the mean fitness in a finite population.

An interesting property of the drift load as first discovered by Robertson (1970a) is that it is independent of the selection coefficients and is approximately equal to $1/(4N_e)$ if a pair of overdominant alleles are segregating in a population of effective size $N_e$. More generally, he showed that the drift load is $(n - 1)/(4N_e)$ if $n$ overdominant alleles are segregating. In deriving these results, it was assumed that overdominance is sufficiently strong so that deviation of gene frequencies by random drift around the equilibrium point is small.

The concept of drift load can also be applied to a polymorphic locus maintained by frequency-dependent selection in which each allele becomes advantageous when rare. Such frequency-dependent selection has long been considered one of the possible mechanisms for maintaining genetic variability in Mendelian populations (cf. Wright 1949, Haldane 1954, Clarke and O'Donald 1964).

In fact, this has been suggested as an important cause for maintaining some pattern polymorphisms in land snails (cf. Clarke 1969). In this case, it is argued that predators develop "searching images" for common phenotypes (genotypes) among prey, while overlooking rarer forms. This type of selection is termed "apostatic selection," and it is possible that such a selection is widespread among animals with respect to "visible" characters.

It is in relation to a non-visible character, however, that much attention has recently been paid to frequency-dependent selection. Namely, this type of selection with selective neutrality at equilibrium was proposed by Kojima and Yarbrough (1967) to explain the prevalence of isozyme polymorphism in natural populations. They suggest that, under this type of selection, the load is non-existent or minimal.

In actual populations, however, gene frequencies are likely to deviate from their equilibrium values, and this will create the drift load. Furthermore, the equilibrium frequencies themselves may not be the optimum frequencies giving the maximum mean fitness of individuals, and this will create an additional load which, following Haldane (1959), may be termed "dysmetric load" (cf. Kimura and Ohta 1970a).

In order to investigate these points, let us consider a pair of alleles $A_1$ and $A_2$ and assume the following simple model of frequency-dependent selection (Table 4.3). In this model, $a$ and $b$ are positive constants ($b > a > 0$), and $W_{12}$ is a multiplier which is positive and independent of $p$. The model is equivalent to the one considered by Wright and Dobzhansky (1946), except that the present model contains a factor $W_{12}$ to show that we consider absolute rather than relative fitnesses.

Under random mating, the mean fitness is

$$\bar{w} = W_{12}\{1 - (a - bp)(1 - 2p)\},$$

and the change of gene frequency per generation by selection is

$$\Delta p = \frac{p(1-p)(a-bp)}{1-(a-bp)(1-2p)}.$$

The equilibrium is attained when $p$ is equal to $\hat{p} = a/b$, and this is stable. At this state, all three genotypes have equal fitnesses of $W_{12}$, with the mean fitness of $\hat{\bar{w}} = W_{12}$.

An important property of this system, in contrast to the overdominant system, is that the gene frequency at stable equilibrium ($\hat{p}$) is not generally the one which gives the

TABLE 4.3. A model of frequency-dependent selection.

| Genotype | $A_1A_1$ | $A_1A_2$ | $A_2A_2$ |
|---|---|---|---|
| Fitness (selective values) | $W_{12}(1 + a - bp)$ | $W_{12}$ | $W_{12}(1 - a + bp)$ |
| Frequency | $p^2$ | $2p(1-p)$ | $(1-p)^2$ |

maximum of $\bar{w}$. In fact, the maximum occurs when $p = 1/4 + a/(2b) \equiv p_{max}$, giving

$$\bar{w}_{max} = W_{12}\left\{1 + \frac{(2a-b)^2}{8b}\right\} \tag{32}$$

Thus, $\bar{w}_{max}$ is larger than $\hat{\bar{w}}$ unless $2a = b$ or $\hat{p} = 0.5$. If we define generally the dysmetric load by

$$L_{dys} = \frac{\bar{w}_{max} - \hat{\bar{w}}}{\bar{w}_{max}}, \tag{33}$$

we have

$$L_{dys} = \frac{(2a-b)^2}{8b + (2a-b)^2} \tag{34}$$

for the present model.

In their analysis of the polymorphism involving ST and CH chromosomes in *Drosophila pseudoobscura,* Wright and Dobzhansky (1946) estimated $a = 0.902$ and $b = 1.288$, so if we substitute these values in the above formula, we get $L_{dys} = 0.0252$ or about 2.5%. As another example of frequency-dependent selection, we may take Kojima and Yarbrough's (1967) study on esterase 6 polymorphism in *Drosophila melanogaster.* If we apply the present model to their experimental results, taking $p$ for the frequency of $F$ allele, we get approximately $a = 0.60$ and $b = 1.58$. Substituting these values for $a$ and $b$ in equation (34), we obtain $L_{dys} = 0.0112$ or about 1.1%.

It may be inferred from these examples that the dysmetric load becomes large when a number of polymorphic loci are maintained by frequency-dependent selection, even if apparent selective neutrality is attained at equilibrium. In other words, an enormous increase of the mean fitness of individuals may be possible by simply adjusting the relative frequencies of alleles at individual loci to their optimum values.

Next, we shall estimate the drift load, assuming that the frequency-dependent selection is sufficiently strong that the gene frequency distribution clusters around the equilibrium value $\hat{p}$. Under this assumption, we may use the approximation

$$\Delta p = -k(p - \hat{p})$$

for the rate of change in gene frequency where $k = \hat{p}(1 - \hat{p})b$ in the present model. Also, we have $E(p) = \hat{p}$, because if $p'$ is the frequency in the next generation, then $p' = p + \Delta p + \xi$, in which $\xi$ is the change due to random sampling with mean 0 and variance $p(1 - p)/(2N_e)$, so that $E(p') = E(p)$ means $E(\Delta p) = 0$.

In order to calculate the drift load, we note that $\bar{w}$ is expressed in terms of $(p - \hat{p})$ as

$$\bar{w} = W_{12}\{1 + (b - 2a)(p - \hat{p}) - 2b(p - \hat{p})^2\}$$

and therefore

$$E(\bar{w}) = W_{12}\{1 - 2bE((p - \hat{p})^2)\}. \tag{35}$$

Applying the definition of the drift load (31), we have

$$L_{\text{drift}} = 2bE\{(p - \hat{p})^2\}. \tag{36}$$

The expected value, $E\{(p - \hat{p})^2\}$, appearing in the right-hand side of this equation may be obtained as follows. Squaring both sides of

$$p' = p + \Delta p + \xi,$$

taking expectation, noting $E(\xi) = 0$ and $E(\xi^2) = p(1 - p)/(2N_e)$ but neglecting the small term $(\Delta p)^2$, we have, at equilibrium, in which $E(p') = E(p)$, the relation

$$E\left\{2p\Delta p + \frac{p(1 - p)}{2N_e}\right\} = 0.$$

If we substitute the approximate expression $\Delta p = -k(p - \hat{p})$

for $\Delta p$ in this equation, we obtain

$$E\{p - \hat{p})^2\} = \frac{\hat{p}(1 - \hat{p})}{4N_e k + 1}, \tag{37}$$

where $k = \hat{p}(1 - \hat{p})b$ in the present model. Since $k$ represents the selective force that pushes $p$ back toward $\hat{p}$ when $p$ deviates from $\hat{p}$, our assumption of small deviation of gene frequency from its equilibrium value means $4N_e k \gg 1$. Thus

$$E\{(p - \hat{p})^2\} = \frac{\hat{p}(1 - \hat{p})}{4N_e k} \tag{38}$$

approximately. Substituting this in the right-hand side of equation (36) and noting $k = \hat{p}(1 - \hat{p})b$ in the present model, we obtain

$$L_{\text{drift}} = \frac{1}{2N_e}. \tag{39}$$

An interesting point to note here is that this load is independent of parameters $a$, $b$ and $W_{12}$. Also, it can be shown that if the absolute fitnesses of $A_1A_1$, $A_1A_2$, $A_2A_2$ are $W_{12}$, $W_{12}(1 - a + bp)$, $W_{12}(1 - 2(a - bp))$ rather than $W_{12}$ $(1 + a - bp)$, $W_{12}$, $W_{12}(1 - a + bp)$ as assumed above, the drift load is unchanged, even if the dysmetric load changes with this new assignment of fitnesses.

These are simple examples of frequency-dependent selection which may be called gene-frequency-dependent selection. Although there are infinite ways of assigning functions of $p$ to selective values of genotypes, we shall consider in what follows a couple of simple models that may be called genotype-frequency-dependent selection. In the first model, we assume that the heterozygotes are distinct and the absolute fitnesses of $A_1A_1$, $A_1A_2$, $A_2A_2$ are respectively $C(1 - sp^2)$, $C[1 - sp(1 - p)]$, $C[1 - s(1 - p)^2]$, where $C$ and $s$ are positive constants. In the second model, we assume that $A_1$ is completely recessive and the fitnesses of the recessive ($A_1A_1$) and the dominants ($A_1A_2$, $A_2A_2$) are respectively $C(1-sp^2)$ and $C[1-s(1-p^2)]$. These two models have been considered by Clarke and O'Donald (1964), who have shown that the equilibrium in either model is stable.

Using the same procedure as used to derive the drift and dysmetric loads for gene-frequency-dependent selection, we can show that in these two models of genotype-frequency-dependent selection the drift load is again $1/(2N_e)$ while dysmetric load is zero. We can modify the second model by assuming fitnesses $C[1-s_a(p^2-a^2)]$ and $C[1-s_A(a^2-p^2)]$ to the recessives and the dominants, where $C$, $s_a$, $s_A$ and $a^2$ are non-negative constants. We can then show that $L_{\text{drift}}=1/(2N_e)$ if $L_{\text{dys}}=0$. More generally, we can show that $L_{\text{drift}}>1/(4N_e)$.

Returning to gene-frequency-dependent selection, let us consider a simple model with multiple alleles. Suppose that the population contains $n$ alleles $A_1, A_2, \ldots, A_n$ and that the absolute fitness of genotype $A_iA_j$ is $C(1-s_ip_i-s_jp_j)$ where the $s_i$'s and $C$ are positive constants. This model was considered earlier by Wright (1949), who showed that there is a stable equilibrium if $s_i>0$ ($i=1, 2, \ldots, n$).

Assuming that the deviation of gene frequencies by random drift from their equilibrium values is small, we can show (Kimura and Ohta 1970a) that $L_{\text{dys}}=0$ and

$$L_{\text{drift}}=\frac{n-1}{2N_e}. \tag{40}$$

For two alleles ($n=2$), this reduces to our previous result (formula 39).

It is interesting to note here that the drift load for the gene-frequency-dependent selection with $n$ alleles is twice as large as the corresponding value for the case of over-dominance (cf. Robertson 1970a).

CHAPTER FIVE

# Adaptive Evolution and Substitutional Load

The process of adaptive evolution, from our standpoint, consists of a sequence of events through which advantageous mutant genes become established in the population. An essential point is that it is a process of gene substitution governed by the action of natural selection rather than by random genetic drift. In Chapter 2, we have presented evidence for a great deal of random substitution of nucleotides, but in this chapter we are concerned with changes that are adaptive. In reality there may be no abrupt change between these two classes of mutants, but it is convenient to treat them as two distinct classes.

In the process of gene substitutions by natural selection, elimination of less fit genes is involved and therefore a genetic load is created, a load which we shall call the substitutional load.

The concept of substitutional load is based on the work of Haldane published in 1957 in a paper entitled "The cost of natural selection." In this paper, he considered a situation in which an original gene becomes disadvantageous due to change of environment, while a mutant allele that was originally less fit becomes advantageous in the new environment, and increases from a low frequency to a high frequency and finally to fixation. He then showed, using a deterministic model, that the total number of selective deaths needed to carry out the gene substitution is almost independent of the selection coefficient. For example, in a haploid organism, if the initial frequency of the mutant allele is $p_0$, the cumulative total of the fraction of selective deaths is $-\log_e p_0$. Thus, if $p_0 = 10^{-5}$, the cost for one gene substitution is about 11.5. In other words, the total number of genetic deaths involved in the whole process is some 11.5 times the population number in one generation. Although spread over many generations, a larger part of the total deaths occur while the advantageous mutant gene is still rare. He also made a similar calculation for diploid organisms. For example, if the gene effect is additive, the corresponding cost is $-2 \log p_0$. If the gene is recessive, the cost may be considerably more. He then took 30 as a typical cost for one gene substitution and suggested that, in horotelic (standard rate) evolution, the rate of substitution is about 1 in every 300 generations. At equilibrium, this gives 10% selective death per generation due to gene substitution. He considered that this accords with the observed slowness of actual evolution.

Since then many papers have been published on the subject, trying to extend the theory to cover the situations in which the selection coefficient is large (Haldane 1960), epistatic interaction in fitness is involved (Kimura and Crow 1969), environmental change is slow (Kimura 1967), the effective population number is not large (Kimura and Maruyama 1969), etc. Also, attempts have been made to apply its concept to various evolutionary problems (Kimura 1960a, 1961a, 1967, 1968a), as well as to clarify the concept itself (Crow 1968a, 1970, Kimura and Crow 1969, Crow and Kimura 1970, Nei 1970, Felsenstein 1971).

However, the concept is still a controversial one, having been criticized by a number of authors. One popular criticism is that the substitution of a more advantageous allele for a less advantageous one cannot be considered a "load," since the population fitness is thereby increased, and therefore the limitation to the evolutionary rate through the cost does not actually exist. This type of criticism, we believe, overlooks the general biological fact that for each species the physical as well as biotic environment is constantly deteriorating (through unpredictable change in climate and especially through evolution of competitors), while advantageous mutants are always very rare at the start.

Recently, Feller (1967) took Haldane's selective death to mean the actual decrement of the number of disadvantageous genes from one generation to the next. While disregarding the population regulating mechanism, he claimed that the cost obtained by his method of calculation may be much less than Haldane's cost, and that the discrepancy comes from Haldane's neglect of the change in population number. Kimura and Crow (1969) pointed out that Feller had misinterpreted Haldane's intention, since what is pertinent here is not the actual numbers of genes but their proportions. They think that Feller overlooked an important biological fact that in the process of gene substitution in evolution, the relative proportions of genes change enormously, while the total population number remains relatively constant due to population regulating mechanisms. Therefore, in the first approximation, the total population number is determined by such factors as food, space and competitors, rather than the relative frequencies of alleles at a particular locus where a substitution is occurring.

It is often not realized how important is the power of natural selection by which a more advantageous gene can gradually supplant less advantageous genes in a population without appreciably affecting the total population number. By this power, an advantageous gene combination that was initially very rare or even non-existent can finally emerge as the prevailing type in a population whose total number is always restricted by the carrying capacity of the environment.

Let us first investigate the problem of gene substitution and the accompanying load using a simple deterministic model. We assume a population of haploid organisms having a large number of loci. Let $K$ be the rate of occurrence per generation of new mutants starting the process of increasing toward eventual replacement of the original allele. We assume that each of these is caused by an environmental change as envisaged by Haldane.

If this process goes on for a sufficiently long time, a steady state will be reached with respect to the frequency distribution of advantageous alleles among different loci. At this statistical equilibrium, each generation, new mutants start their journey toward fixation on the average at $K$ of the loci, while fixation is just completed also at the same number of loci. Thus $K$ represents the rate of gene substitution in the population per generation, being about $1/300$ according to Haldane's conjecture.

In order to simplify the treatment, we assume that the selective advantages of mutants as well as their initial frequencies are the same among different loci. We also assume that fitness ($w$) measured in selective values is multiplicative between different loci so that log $w$ is additive between loci. With a large number of independently segregating loci, log $w$ will be distributed approximately normally. Consider a particular locus in which a mutant gene $A_1$ has started to increase. Let $s_1$ be the selective advantage of $A_1$ over its allele $A_2$ such that the relative fitnesses (in selective values) of $A_1$ and $A_2$ are 1 and $1 - s_1$ respectively. More precisely, we may assume that the selective advantage per short time interval of length $\Delta t$ is $s_1 \Delta t$. Then, using a continuous approximation, we have

$$\frac{dp}{dt} = s_1 p(1 - p), \tag{1}$$

where $p$ stands for the frequency of mutant gene $A_1$. Applying this equation to all the loci at which gene substitution is going on, we find that, since $K\Delta t$ mutants start to increase during any interval of length $\Delta t$, the number of loci in which the frequency of the advantageous mutant is in the range $p \sim p + dp$ is given by

$$\Phi(p)\, dp = \frac{K\, dp}{s_1 p(1-p)}, \tag{2}$$

where $p_0 < p < 1$. This is a deterministic version of the corresponding distribution which we shall derive later in this section based on a stochastic model.

Now consider the load. At a particular locus where the frequency of the advantageous mutant is $p$, the mean fitness of the population, in $\log w$ scale, is less by $-(1-p)\log(1-s_1)$ or approximately $s_1(1-p)$ than the fitness of the favored type $A_1$ (Table 5.1). Thus, summing over all the relevant loci, the substitutional or evolutional load, in $\log w$ scale, is

$$L_e = \int_{p_0}^{1} s_1(1-p)\Phi(p)\, dp$$

$$= \int_{p_0}^{1} \frac{Ks_1(1-p)\, dp}{s_1 p(1-p)} = -K \log p_0 \tag{3}$$

This is the load at a steady state when gene substitution proceeds at the uniform rate $K$ per generation. Note that this load is independent of the selection coefficient $s_1$. Furthermore, $s_1$ ($>0$) need not be constant throughout the process of substitution, because $s_1$ in the numerator and in

TABLE 5.1. Table of fitnesses in the haploid model.

| Genotype | $A_1$ | $A_2$ |
|---|---|---|
| Frequency | $p$ | $1-p$ |
| Fitness | | |
| $w$ | 1 | $1-s_1$ |
| $\log w$ | 0 | $\log(1-s_1) \approx -s_1$ |

the denominator in the above integral cancel each other in each infinitesimal time interval.

The total number of loci at which gene substitution is going on may be estimated by

$$n_L = \int_{p_0}^{1-1/N} \Phi(p)\,\mathrm{d}p = \frac{K}{s_1}\int_{p_0}^{1-1/N}\left(\frac{1}{p}+\frac{1}{1-p}\right)\mathrm{d}p.$$

If we assume $p_0 = 1/N$, this reduces approximately to

$$n_L = (2K/s_1)(\gamma + \log_e N) \tag{4}$$

where $\gamma \approx 0.577$.

It may be of some interest to compute the variance in fitness among individuals. Since the variance is $p(1-p)$ $[\log(1-s_1)]^2$ or approximately $s_1{}^2 p(1-p)$ at a locus with gene frequency $p$, summing over all the relevant loci, we have

$$\begin{aligned} Vr\,(\log w) &= \int_{p_0}^{1} s_1{}^2 p(1-p)\Phi(p)\,\mathrm{d}p \\ &= \int_{p_0}^{1} \frac{Ks_1{}^2 p(1-p)}{s_1 p(1-p)}\,\mathrm{d}p \\ &= Ks_1(1-p_0) \approx Ks_1. \end{aligned} \tag{5}$$

We will denote the square root of this by $\sigma\,(\log w)$.

Noting that $\delta\,(\log w) \approx \delta w/w$, we may express this result also as

$$\frac{Vr(w)}{\bar{w}^2} = Ks_1 \tag{6}$$

where $\bar{w}$ and $Vr(w)$ are respectively the mean and the variance of fitness in selective values. An equivalent result using the concept of "cost" has been obtained by Crow (1970) and also by Crow and Kimura (1970). If the rate of gene substitution is expressed in terms of one substitution every $n$ generations, then $K$ in the above formulae may be replaced by $1/n$.

Essentially the same treatment can be applied to a diploid population if the mutant allele ($A_1$) is semidominant. Namely, if $s_1$ is the selective advantage of $A_1$ over $A_2$ so that the selective advantage of $A_1A_1$ over $A_2A_2$ is $2s_1$, the formula for the frequency distribution of mutants among different loci is unchanged and is given by (2), while $L_e$ and $Vr$ (log $w$) become twice as large. Namely,

$$L_e = -2K \log_e p_0 \tag{7}$$

and

$$Vr\ (\log w) \approx 2Ks_1. \tag{8}$$

The total number of loci at which the substitution is going on may be estimated in the same way as in (5) to give $n_L = (2K/s_1)(\gamma + \log_e 2N)$. The number of heterozygous loci per individual is

$$H = \int_{1/(2N)}^{1-1/(2N)} 2p(1-p)\Phi(p)\,dp \approx \frac{2K}{s_1}. \tag{9}$$

An interesting point to note is that $L_e$ is independent of $s_1$ but depends on $p_0$, and becomes large if $p_0$ is small. On the other hand, $Vr$ (log $w$) depends directly on $s_1$ but is practically independent of $p_0$ if it is small.

The substitutional load given in (3) represents the average depression of the mean fitness relative to the optimum genotype having fixed all the advantageous mutant genes currently segregating in the population. For example, if $K = 1, p = 10^{-5}$, we have $L_e = 11.5$, which means that mean number of offspring is only $e^{-11.5}$ of that of the optimum genotype. The number of segregating loci is approximately $n_L = 2{,}400$.

If gene substitution is going on simultaneously at a large number of loci, as in the above haploid example, such an optimum genotype is unlikely to appear in the population. Nevertheless, the load thus calculated has a realistic meaning in a changing environment. For example, an individual may be killed because it does not have mutant genes giving resistance to new infectious diseases, newly evolved predators, unusually severe cold, etc. In such a situation, the probability of death of an individual may be influenced little by whether the multi-mutant individual occurs in the population or not. The deterioration of the environment is not compensated until multi-mutant individuals become common. Such a situation must occur very often in adaptive evolution. The population is always behind in the struggle to keep its genotypic frequencies adjusted to a continuously changing environment. As Felsenstein (1971) writes, if a population could substitute beneficial genes instantly after a change in environment there would be no cost; natural selection has a cost in the sense that it is less efficient than divine intervention.

On the other hand, mutant genes may spread into the population simply because they enhance the competitive ability of individuals within the population. In this case, the load given by $L_e$ may overestimate the actual amount of selective elimination, and it may be more appropriate to consider "the substitutional load based on competition" (Kimura 1969b). Environmental change in the usual sense need not be involved in such gene substitution. It is likely, however, that such a substitution is often detrimental to the species and may have an effect equivalent to environmental deterioration, which has to be compensated further by additional gene substitutions in order that the species can survive in its struggle against competing species. We must keep in mind that in nature each species is constantly improving its fitness by gene substitution, yet it is not getting better off in most of the cases simply because its competitors are improving their fitness at the same time.

Let us suppose for the moment that the gene substitutions are being carried out solely based on intraspecies competition. In this situation, the variance of fitness as given by (6) may be more appropriate to compute the load than the mean depression of fitness relative to the optimum genotype. Namely, at statistical equilibrium in which gene substitution is proceeding at the rate $K$ per generation, the difference of fitness in log $w$ scale between the mean and the individual having the most probable highest fitness within the population is

$$\tilde{x}_{N,1} \approx \sqrt{2 \log_e (0.4N)}$$

times $\sigma$ (log $w$) $= \sqrt{Ks_1}$, where $N$ is the number of individuals in the population. The above asymptotic formula for $\tilde{x}_{N,1}$ is Frank's formula giving "the most probable largest normal value" (cf. Gumbel 1958). Then the substitutional load measured in log $w$ scale with respect to competitive ability is given by

$$\tilde{L}_e = \tilde{x}_{N,1}\sigma (\log w)$$

or

$$\tilde{L}_e = \sqrt{2Ks_1 \log_e (0.4N)}. \qquad (10)$$

For example, if $K = 1$, $N = 10^5$, $s_1 = 0.01$, we have

$$\tilde{L}_e = \sqrt{0.02 \times \log_e (4 \times 10^4)} = 0.46.$$

This means that, disregarding the environmental effect on fitness, the individual carrying the largest number of advantageous mutants in the population must have about $e^{\tilde{L}_e} = 1.58$ times as many offspring as the average individual.

Generally, $\tilde{L}_e$ is smaller than $L_e$, and indeed, $\tilde{L}_e$ can be made as small as one pleases by making $s_1$ small. The above formula for $\tilde{L}_e$ was derived assuming a haploid population, but the same formula holds for a diploid population if the effect of gene substitution at each locus is additive, provided that $s_1$ is replaced by $s$ where $s$ represents the selective advantage of the mutant homozygote.

We shall now treat the problem using a stochastic model. This is more satisfactory than the deterministic model, since the advantageous mutants are usually very rare at the start so that their fate in early generations must be largely controlled by random sampling of gametes, especially when the selection coefficient is small.

We shall use the method of diffusion equations, and in particular, we shall apply the theory developed by Kimura (1969b), who derived the gene frequency distribution attained under a steady flux of mutations in a finite population. Although the theory was originally developed in terms of nucleotide sites, it can be applied to a conventional genetic locus if the fraction of segregating loci due to gene substitution is small.

We consider a Mendelian population consisting of $N$ diploid individuals and having "variance" effective number $N_e$.

Let us assume that a very large number of loci are available for possible gene substitution, and that, in the entire population, in each generation mutant genes acquire a selective advantage on the average at $\nu_m$ of the loci. We also assume that the number of loci is large enough that each of these events takes place at a different locus.

Consider a particular locus at which an advantageous mutant allele has started to increase. Let $p$ be its initial frequency. If the mutant allele is advantageous from the moment of its birth and if every mutant represents a new, not preexisting allele, $p = 1/(2N)$, where $N$ is the number of individuals in the population. On the other hand, if the mutant allele is recurrent and initially disadvantageous but becomes advantageous later due to change of environment, $p$ can be larger than $1/(2N)$.

Let $\phi(p, x; t)$ be the probability density that the frequency of mutant allele in the population becomes $x$ after $t$ generations, given that it is $p$ at $t = 0$.

Note that in the stochastic treatment we use the letter $x$ as a random variable representing gene frequency and use $p$ as a constant representing initial frequency. In the deterministic treatment, on the other hand, we have used the letter $p$ to represent gene frequency at an arbitrary generation and $p_0$ to represent the initial frequency.

Our main interest is the distribution of the frequencies of the advantageous mutant alleles among different loci at statistical equilibrium in which gene substitution is proceeding at a constant rate.

Let $\Phi(p, x)\,dx$ be the expected number of loci in which the frequencies of the mutant allele lie between $x$ and $x + dx$. We assume that the initial frequencies and the selective advantages of mutants are the same among different loci. Since the mutant genes start to increase in $\nu_m$ of the loci in the population each generation, and since $\phi(p, x; t)$ is the probability density of the mutant reaching frequency $x$ after $t$ generations, $\nu_m \phi(p, x; t)\,dx$ represents the contribution made by mutants which appeared $t$ generations earlier to the present frequency class having mutant frequency $x \sim x + dx$. Therefore, considering all the contributions made by mutant genes in the past, we have, at statistical equilibrium,

$$\Phi(x, p) = \nu_m \int_0^\infty \phi(x, p; t)\,dt \tag{11}$$

where $0 < x < 1$. An important property of this distribution function is that various statistical parameters such as the numbers of segregating and heterozygous loci, the mean and the variance in fitness as well as the loads can be expressed by integrating this function with respect to $x$ after multiplying it by a suitable function, $f(x)$. Thus, we consider a general expression (functional)

$$I_f(p) = \int_0^1 f(x)\Phi(p, x)\, \mathrm{d}x,$$

where the integral is over the open interval (0,1).

It may be more appropriate to use $1/(2N)$ and $1 - 1/(2N)$ as the limits of integration, especially if the value of the integral is changed significantly by including $x = 0$ and 1.

It can then be shown (Kimura 1969b) that this satisfies the differential equation

$$\tfrac{1}{2} V_{\delta p} I_f''(p) + M_{\delta p} I_f'(p) + \nu_m f(p) = 0 \tag{12}$$

with boundary conditions

$$I_f(0) = I_f(1) = 0 \cdot$$

In this equation, $M_{\delta p}$ and $V_{\delta p}$ are respectively the mean and the variance of the change in mutant frequency per generation. The solution of this equation is

$$\begin{aligned} I_f(p) = \{1 - u(p)\} \int_0^p \psi_f(\xi) u(\xi)\, \mathrm{d}\xi \\ + u(p) \int_p^1 \psi_f(\xi)\{1 - u(\xi)\}\, \mathrm{d}\xi, \end{aligned} \tag{13}$$

where

$$\psi_f(\xi) = \frac{2\nu_m f(\xi) \int_0^1 G(x)\, \mathrm{d}x}{V_{\delta\xi} G(\xi)}$$

depends on $f(\xi)$, and

$$u(p) = \frac{\int_0^p G(x)\, \mathrm{d}x}{\int_0^1 G(x)\, \mathrm{d}x}$$

is the probability of eventual fixation in which

$$G(x) = \exp \{-2 \int_0^x \frac{M_{\delta\xi}}{V_{\delta\xi}}\, d\xi\}.$$

In the following, we shall consider a more restricted case in which the selective advantage of the mutant homozygotes and heterozygotes in each locus are given respectively by $s$ and $sh$ so that, under random mating,

$$M_{\delta p} = sp(1-p)\{h + p(1-2h)\}.$$

More precisely, if we assign selective values $1 + s\Delta t$, $1 + sh\Delta t$ and 1 respectively to mutant homo-, hetero- and normal homozygotes for a short time interval of length $\Delta t$, we have the expression in the right-hand side for $M_{\delta p}/\Delta t$ as $\Delta t \to 0$.

We also assume that the "variance" effective size of the population is $N_e$ so that

$$V_{\delta p} = \frac{p(1-p)}{2N_e}.$$

Then, at a particular locus in which the mutant frequency is $x$, the mean fitness in the $\log w$ scale is less by $\log(1+s) - \log[1 + sx^2 + 2shx(1-x)]$ or approximately $s - sx^2 - 2shx \times (1-x)$ than that of the mutant homozygote (assuming $s > 0$ and $1 > h > 0$).

This means that the substitutional load is given by $I_f(p)$ with $f(x) = s(1-x)\{1 + (1-2h)x\}$, where $f(x)$ represents the load at a locus having mutant frequency $x$. Substituting this expression for $f$ in (13), we obtain, after some rearrangements,

$$L_e = KL(p), \tag{14}$$

where

$$K = \nu_m u(p) \tag{15}$$

is the rate of gene substitution per generation, and

$$L(p) = 4S\left\{\frac{1}{u(p)} - 1\right\}\int_0^1 A(x)\,\mathrm{d}x\int_0^x G(\lambda)\,\mathrm{d}\lambda$$

$$-\frac{4S}{u(p)}\int_p^1 A(x)\,\mathrm{d}x\int_p^x G(\lambda)\,\mathrm{d}\lambda \qquad (16)$$

is the load (or cost) of one gene substitution. These agree with the results obtained by Kimura and Maruyama (1969). In the above formulae, $S$, $G(x)$ and $A(x)$ are given by

$$S = N_e s,$$

$$G(x) = \exp\{-4Shx - 2S(1-2h)x^2\}$$

and

$$A(x) = \frac{(1/x) + (1-2h)}{G(x)}.$$

In the simplest but important case in which the effect of mutant is additive within a locus (i.e. the case of semidominance or "no dominance"), we may put $h = 1/2$ in the above formulae. Then the expression for the load is simplified to give

$$L(p) = 2\left\{\frac{1}{u(p)} - 1\right\}\int_0^{2Sp}\frac{e^y - 1}{y}\,\mathrm{d}y$$

$$- 2e^{-2S}\int_{2Sp}^{2S}\frac{e^y}{y}\,\mathrm{d}y - 2\log_e p, \qquad (17)$$

where

$$u(p) = \frac{1 - e^{-2Sp}}{1 - e^{-2S}}, \qquad (18)$$

in which $S = N_e s$.

There are two cases of particular interest. First, if the selective advantage of the mutant is large enough so that $2N_e s$ is much larger than unity while the initial frequency of the mutant is so low that $2N_e sp$ is much smaller than unity, we have approximately

$$L(p) = -2 \log_e p + 2$$

and

$$u(p) = 2N_e sp.$$

In this case, the load for one gene substitution (cost) is larger by about two than the corresponding value first obtained by Haldane using a deterministic model. This is due to the fact that some of the advantageous mutations are lost by chance, never contributing to substitution, and their contribution to the load is wasted. If each mutant represents a new, not preexisting allele at a different locus, $p = 1/(2N)$. Then, we have $K = (N_e/N)s\nu_m$ for the rate of gene substitution. If we denote by $s_1$ the effect of a mutant in single dose so that $s_1 = s/2$, we have

$$K = \frac{2N_e}{N} s_1 \nu_m. \qquad (19)$$

In the second case, if the mutant gene is almost neutral to selection such that $|2N_e s| \ll 1$, we have approximately

$$L(p) = -4N_e s \log_e p \qquad (20)$$

and

$$u(p) = p + N_e sp(1 - p). \qquad (21)$$

In this case, as $N_e s$ approaches zero, the probability of fixation approaches $p$ and the substitutional load may become indefinitely small. For such mutations, there will be no limit to the rate of gene substitution imposed by the substitutional load. The limiting factor is the rate by which such mutations are produced. As discussed in Chapter 2, a majority of mutant substitutions in molecular evolution appear to belong to this category.

The functional $I_f(p)$ considered above is quite general

and we can obtain, for example, the number of segregating loci in the population by putting $f(x) = 1$, and the average number of heterozygous loci per individual ($H(p)$) by putting $f(x) = 2x(1 - x)$. In the special case of semidominance in fitness, if $2N_e s \gg 1$, it can be shown (Kimura 1969b) that the average number of heterozygous loci per individual is approximately

$$H(1/2N) \approx 4v_m \frac{N_e}{N}. \tag{22}$$

This agrees with (9) if we note that $K = (N_e/N)sv_m$ and $s_1 = s/2$. If, on the other hand, the mutant is almost neutral

$$H(1/2N) \approx 2v_m \frac{N_e}{N}. \tag{23}$$

Equation (22) suggests that mutations having a definite advantage cannot contribute greatly to the heterozygosity of an individual unless a large number of them appear in the population each generation.

The functional $I_f(p)$ can also be used to obtain the variance of fitness among individuals in log $w$ scale by putting $f(x) = s^2x(1 - x)\{2h^2 + (1 - 4h^2)x + (1 - 2h)^2x^2\}$, which is the genotypic variance in fitness at a locus with mutant frequency $x$. For semidominant mutations, $h = 1/2$ and we have $f(x) = 2s_1^2x(1 - x)$ so that $Vr(\log w) = s_1^2H(p)$. Thus, if the mutations are definitely advantageous ($2N_e s \gg 1$), we have

$$Vr(\log w) \approx 4s_1^2 v_m \frac{N_e}{N},$$

or noting $K = (2N_e/N)s_1 v_m$, we have approximately

$$Vr(\log w) = 2Ks_1,$$

which agrees with (8) obtained using a deterministic model.

If the mutations are almost neutral this can of course be

indefinitely small. In this case, if we take each nucleotide site rather than the conventional genetic locus as the unit of mutation, equation (23) represents the number of heterozygous nucleotide sites per individual when $\nu_m$ molecular mutations appear in the entire population each generation. Let $v_0 = \nu_m/(2N)$ be the rate per gamete per generation for neutral mutations, the average number of heterozygous sites per individual due to such mutations is

$$H = 4N_e v_0. \tag{24}$$

For example, in a mammalian population with the effective size of 25,000, if the neutral mutations appear at the rate of 10 per gamete per generation, an average individual in a population is heterozygous at a million nucleotide sites. If this appears to be large, it nevertheless amounts only to about 0.03% of the nucleotides making up the genome.

CHAPTER SIX

# Two-locus Problems

Just as two-body problems are more difficult than single-body problems in physics, the treatment of two-locus behavior of genes is more intricate and difficult than that of single-locus behavior.

In his pioneering work, Haldane (1931b) investigated the course of change in gene frequencies under epistatic (non-additive) interaction in fitness between two loci. In his treatment, he assumed completely random association of genes, that is, he neglected the complication due to linkage. His treatment is still a good approximation if linkage is loose and epistatic interaction is weak. On the other hand, Wright (1945b) investigated the departure from random combination of genes using a simple optimum model with quadratic interaction in fitness. In his treatment, Wright left out the problem of the stability of equilibrium. Similarly, Fisher (1930b), in his discussion on the evolutionary influence of gene interaction on diminishing recombination did not go into the stability problem.

The general equations for the rate of change in chromosome frequencies when two alleles are segregating in each locus were first given by Kimura (1956a) assuming a population with continuous births and deaths. He then investigated a model of polymorphism suggested by P. M. Sheppard and showed that if the first locus is maintained in balanced polymorphism by overdominance and if there is an epistatic interaction, as will be discussed later, the second locus remains polymorphic if linkage is sufficiently close.

The corresponding equations for a population with discrete generations were obtained by Lewontin and Kojima (1960).

Let us consider a large random mating population of diploid organisms having discrete generations and assume that a pair of alleles $A_1$ and $A_2$ are segregating at the first locus, and alleles $B_1$ and $B_2$ at the second locus. Let $X_1$, $X_2$, $X_3$ and $X_4$ be respectively the frequencies of four chromosome types $A_1B_1$, $A_1B_2$, $A_2B_1$ and $A_2B_2$ immediately after fertilization. It may be convenient here to give these chromosomes the numbers 1, 2, 3 and 4 so that the frequency of chromosome $i$ is $X_i$ and fitness in selective values of the genotype formed by the union of chromosomes $i$ and $j$ may be denoted by $w_{ij}$, where $i, j = 1, 2, 3, 4$ and $\sum_i X_i = 1$.

Let $c$ be the recombination fraction between the two loci, then the rates of change per generation of these chromosome frequencies are

$$\left.\begin{aligned} \Delta X_1 &= \{X_1(w_{1\cdot} - \bar{w}) - cD_w\}/\bar{w} \\ \Delta X_2 &= \{X_2(w_{2\cdot} - \bar{w}) + cD_w\}/\bar{w} \\ \Delta X_3 &= \{X_3(w_{3\cdot} - \bar{w}) + cD_w\}/\bar{w} \\ \Delta X_4 &= \{X_4(w_{4\cdot} - \bar{w}) - cD_w\}/\bar{w} \end{aligned}\right\} \tag{1}$$

where

$$w_{i\cdot} = \sum_{j=1}^{4} w_{ij}X_j,$$

$$\bar{w} = \sum_{ij} w_{ij}X_iX_j = \sum_i w_{i\cdot}X_i$$

and

$$D_w = w_{14}X_1X_4 - w_{23}X_2X_3.$$

In the above expressions, $w_{i\cdot}$ is the average selective value of chromosome $i$, $\bar{w}$ is the average fitness of the population and $D_w$ represents the amount of linkage disequilibrium after selection. By "linkage disequilibrium" we mean non-random association of genes between different loci.

The set of equations (1) can be derived as follows. Consider, for example, chromosome 1. At reproduction, the gamete containing this chromosome ($A_1B_1$) may be derived either from one of the genotypes carrying this chromosome or from genotype $A_1B_2/A_2B_1$ through recombination.

In the former case, the fraction of gametes containing this chromosome is 1 if the genotype is the homozygote $A_1B_1/A_1B_1$, is 1/2 if the genotype is one of the single heterozygotes $A_1B_1/A_2B_1$ or $A_1B_1/A_1B_2$, and is $(1-c)/2$ if the genotype is coupling double heterozygote $A_1B_1/A_2B_2$, which contributes this chromosome only in non-recombinant gametes. On the other hand, with genotype $A_1B_2/A_2B_1$, the fraction of gametes containing this chromosome is $c/2$, because this genotype (repulsion double heterozygote) can produce $A_1B_1$ only through recombination, which occurs with probability $c$.

Thus, multiplying these fractions by the corresponding frequencies of the genotypes after selection, the frequency of chromosone 1 in the next generation is

$$X_1' = \frac{1}{\bar{w}}\{X_1^2 w_{11} + \frac{1}{2}(2X_1X_2w_{12} + 2X_1X_3w_{13}) + \frac{1-c}{2}(2X_1X_4w_{14}) + \frac{c}{2}(2X_2X_3w_{23})\},$$

where the normalizing factor $\bar{w}$ is introduced in the denominator to make the sum of the frequencies of four chromosome types equal to unity.

Noting that $w_{1\cdot} = w_{11}X_1 + w_{12}X_2 + w_{13}X_3 + w_{14}X_4$ and $D_w = w_{14}X_1X_4 - w_{23}X_2X_3$, the above formula for $X_1'$ becomes

$$X_1' = (X_1w_{1\cdot} - cD_w)/\bar{w},$$

from which we obtain

$$\Delta X_1 = X_1' - X_1 = \{X_1(w_{1\cdot} - \bar{w}) - cD_w\}/\bar{w},$$

the first equation in (1). The remaining equations may be obtained similarly by considering $X_2'$, $X_3'$ and $X_4'$. This set of equations is slightly more general than the ones given by Lewontin and Kojima (1960) but equivalent to those of Bodmer and Parsons (1962).

This method of derivation may readily be extended to treat the multi-allelic case in which $n$ alleles $A_1, A_2, \ldots, A_n$ are segregating at the first locus and $m$ alleles $B_1, B_2, \ldots, B_m$ are segregating at the second locus. In such a case, it is convenient to use double subscripts to represent chromosome frequencies so that $X_{ij}$ stands for the frequency of chromosome $A_iB_j$. We shall denote the fitness in selective values of genotype $A_iB_j/A_kB_\ell$ by $w_{ij,k\ell}$ $(= w_{k\ell,ij})$. Applying the same process of reasoning as was used to derive $X_1'$ in the two-allelic case, we obtain, assuming random mating, the frequency of chromosome $A_1B_1$ in the next generation,

$$X_{11}' = \frac{1}{\bar{w}}\{w_{11,11}X_{11}^2 + \sum_i{}' w_{11,i1}X_{11}X_{i1} + \sum_j{}' w_{11,1j}X_{11}X_{1j}$$

$$+ (1-c)\sum_{i,j}{}' w_{11,ij}X_{11}X_{ij} + c\sum_{i,j}{}' w_{i1,1j}X_{i1}X_{1j}\},$$

where the primed summation is over all values of $i$ and $j$ except for $i = 1$ and $j = 1$. Thus, writing

$$w_{11\cdot\cdot} = \sum_{i=1}^{n}\sum_{j=1}^{m} w_{11,ij}X_{ij}$$

and

$$D_w(11, ij) = w_{11,ij}X_{11}X_{ij} - w_{i1,1j}X_{i1}X_{1j},$$

we obtain the equation for $A_1B_1$,

$$\Delta X_{11} = \{X_{11}(w_{11\cdot\cdot} - \bar{w}) - c\sum_{i,j} D_w(11, ij)\}/\bar{w}, \qquad (2)$$

where

$$\bar{w} = \sum_{i,j} w_{ij\cdot\cdot} X_{ij} = \sum_{i,j,k,\ell} w_{ij,k\ell} X_{ij} X_{k\ell}.$$

The corresponding equations for other chromosomes may readily be obtained in the same way.

If we assume that the fitnesses of the coupling and repulsion heterozygotes are the same, $w_{11,ij} = w_{i1,1j}$ and $D_w$ in equation (2) is

$$D_w(11, ij) = w_{11,ij}(X_{11}X_{ij} - X_{i1}X_{1j}). \quad (3)$$

Such an assumption is realistic in almost all cases when the two loci represent distinct genes. However, if the two loci in the above formulation actually represent two nucleotide sites within a cistron (for which $n \leqq 4, m \leqq 4$), this assumption may not hold because of cis-trans position effects.

Returning to the simpler case of two alleles segregating at each locus, we shall consider the problem of non-random association of genes (usually called "linkage disequilibrium") in an infinitely large population. To simplify the treatment, we shall assume random mating, constant selective value of each genotype and no cis-trans effects in fitness. Even with these restrictions, the problem is still very difficult and no general theories have been developed to treat two-locus behavior, except for the theory relating to "quasi-linkage equilibrium" to be discussed later.

So far, the completely worked out cases are rather special ones involving equilibria, and we shall summarize the main results.

Let

$$D = X_1X_4 - X_2X_3, \quad (4)$$

then

$$D_w = w_{14}D$$

in equations (1). The value $D$ is equal to half the difference between the frequencies of coupling and repulsion heterozygotes, and it becomes zero if alleles at the two loci are in random combination. Thus, $D$ serves as an index to represent the amount of linkage disequilibrium. Note, however, that $D$ also depends strongly on the gene frequencies at each locus.

It can be shown that if the fitnesses are purely additive between loci (i.e. no epistasis), $D$ becomes zero at equilibrium. Non-randomness arises under epistasis, including multiplicative overdominance, as will be shown below. Generally the equilibrium points are determined by the balance between epistatic interaction and recombination, since, by recombination, favorable gene combinations are broken up and the equilibrium is stable if the linkage is sufficiently tight in comparison with the epistatic force. We list several specific models together with the main results.

*Model 1*

We assume that, at the first locus, alleles $A_1$ and $A_2$ are kept in balanced polymorphism by overdominance. At the second locus, alleles $B_1$ and $B_2$ interact with $A_1$ and $A_2$ in such a way that $A_1$ is advantageous in combination with $B_1$ but is disadvantageous in combination with $B_2$, while the situation is reversed for the gene $A_2$. The problem is whether polymorphism is kept at the $B$ locus by the overdominance at the $A$ locus. More specifically, we assume the following table of selective values,

| | $A_1A_1$ | $A_1A_2$ | $A_2A_2$ |
|---|---|---|---|
| $B_1B_1$ | $1+s$ | $1+t$ | $1-s$ |
| $B_1B_2$ | $1$ | $1+t$ | $1$ |
| $B_2B_2$ | $1-s$ | $1+t$ | $1+s$ |

where $0 < s < t$. This model was investigated by Kimura (1956a). Let $\hat{X}_1$, $\hat{X}_2$, $\hat{X}_3$, $\hat{X}_4$ and $\hat{D}$ be the equilibrium chromosome frequencies and linkage disequilibrium. Then, by equating the rates of change given in equation (1) to zero, we get

$$\hat{X}_1 = \hat{X}_4 = 1/2(1/2 - \beta + \sqrt{1/4 + \beta^2})$$
$$\hat{X}_2 = \hat{X}_3 = 1/2(1/2 + \beta - \sqrt{1/4 + \beta^2})$$

and

$$\hat{D} = 1/2(\sqrt{1/4 + \beta^2} - \beta) \quad (5)$$

where $\beta = (1 + t)c/s$. At this equilibrium, the gene frequency at each locus is 1/2. When $\beta = 0$, i.e. $c = 0$, $\hat{D} = 1/4$ and there are only two types of chromosomes, $A_1B_1$ and $A_2B_2$, in the population. The two loci behave like a single locus. As $c$ gets larger, $\hat{D}$ becomes smaller and the association of $B_1$ to $A_1$ and $B_2$ to $A_2$ becomes less strong. For a given value of $t$, the equilibrium point is determined by the ratio $c/s$. It has been shown by Kimura (1956a) that this equilibrium is stable only when $c$ is sufficiently small so that

$$c < \frac{t^2 - s^2}{4t(1 + t)}. \quad (6)$$

*Model 2*

The fitnesses (selective values) of nine genotypes are:

| | $A_1A_1$ | $A_1A_2$ | $A_2A_2$ |
|---|---|---|---|
| $B_1B_1$ | $(1 - s)(1 - t)$ | $1 - t$ | $(1 - s)(1 - t)$ |
| $B_1B_2$ | $1 - s$ | $1$ | $1 - s$ |
| $B_2B_2$ | $(1 - s)(1 - t)$ | $1 - t$ | $(1 - s)(1 - t)$ |

This is a model of multiplicative symmetric overdominance in which $s$ is the selection coefficient against either homozygote at the first locus and $t$ is that at the second locus. It involves epistatic interaction since any deviation from additivity between loci may be called the epistatic interaction. In this case the interaction comes from the term

$st$, where $1 > s, t > 0$. This causes "linkage disequilibrium" at equilibrium. We note that as long as the two loci are additive in fitness the fitness of an individual is a function solely of the fitness contributions of the two loci, so there is no selection to counteract the tendency of recombination to produce linkage equilibrium.

With multiplicative overdominance, however, $\bar{w}$ is maximum when only complementary types of chromosomes $A_1B_1$ and $A_2B_2$ or $A_1B_2$ and $A_2B_1$ exist in the population, and the equilibrium point is determined by the balance between the epistatic interaction and recombination. There are three equilibrium points corresponding to

$$\hat{X}_1 = \frac{1}{4}\left(1 + \sqrt{1 - \frac{4c}{st}}\right),$$

$$\hat{X}_1 = \frac{1}{4}\left(1 - \sqrt{1 - \frac{4c}{st}}\right)$$

and

$$\hat{X}_1 = 1/4. \tag{7}$$

For each of these values of $\hat{X}_1$, we have $\hat{X}_2 = \hat{X}_3 = 1/2 - \hat{X}_1$ and $\hat{X}_4 = \hat{X}_1$. The gene frequency at each locus is always 1/2. The first two equilibrium points with $\hat{D} = \pm 1/4 \sqrt{1 - (4c/st)}$ are stable only when $c < st/4$; otherwise, the system will go to the point with $\hat{D} = 0$. More generally, Bodmer and Felsenstein (1967) have shown, using a multiplicative non-symmetric overdominance model, that for random association with $D = 0$ to be stable, we must have

$$c > \left(\frac{s_1 s_2}{s_1 + s_2}\right)\left(\frac{t_1 t_2}{t_1 + t_2}\right) \tag{8}$$

where $s_1$ and $s_2$ are selection coefficients against two homozygotes ($A_1A_1$ and $A_2A_2$) in the first locus and $t_1$ and $t_2$ are those against $B_1B_1$ and $B_2B_2$ in the second locus. In other words, two multiplicative overdominant loci cannot be polymorphic and at the same time in linkage equilibrium ($\hat{D} = 0$) if the recombination fraction is less than the product of the marginal segregational loads.

*Model 3*

The table of selective values is:

| | $A_1A_1$ | $A_1A_2$ | $A_2A_2$ |
|---|---|---|---|
| $B_1B_1$ | $1 - s$ | $1 - t$ | $1 - s$ |
| $B_1B_2$ | $1 - u$ | $1$ | $1 - u$ |
| $B_2B_2$ | $1 - s$ | $1 - t$ | $1 - s$ |

This model was first studied by Lewontin and Kojima (1960), followed by more complete study by Ewens (1968). There is an epistatic interaction in fitness if $t + u - s \neq 0$. Because of symmetry, the equilibrium gene frequency at each locus is 1/2. As before, under epistatic interaction and sufficiently tight linkage, stable equilibria with $\hat{D} \neq 0$ exist. They are,

$$\hat{X}_1 = \hat{X}_4 = 1/4 \pm \frac{1}{4}\sqrt{1 - \frac{4c}{t + u - s}}, \quad \hat{X}_2 = \hat{X}_3 = 1/2 - \hat{X}_1$$

with

$$\hat{D} = \pm\sqrt{1 - \frac{4c}{t + u - s}}\,. \tag{9}$$

Such equilibria with $\hat{D} \neq 0$ exist only when $c < 1/4(t + u - s)$. Otherwise $\hat{D}$ is 0 with equal frequencies of the four chromosome types at stable equilibrium ($\hat{X}_1 = \hat{X}_2 = \hat{X}_3 = \hat{X}_4 = 1/4$), provided that $s > |t - u|$.

It is generally true that for an extremely tight linkage (i.e. $c = 0$) the equilibria with $\hat{D} \neq 0$ are stable. However, even if $c < 1/4(t + u - s)$, the equilibrium with $\hat{D} \neq 0$ may not be stable. Ewens (1968, 1969) has revealed quite complex linkage behavior in such a case. He has shown that for some sets of selection parameters, the stability region for $c$ is divided into two disjoint intervals, one for small values of $c$ and the other for relatively large values of $c$. No stable equilibria exist for values of $c$ between these two intervals. For example, if $t = 0.78$, $u = 0.82$ and $s = 0.1$, the stability region for $c$ consists of two intervals (0, 0.1005) and (0.3731, 0.375).

*Model 4*

A more general model than the preceding ones may be expressed in the form:

| | $A_1A_1$ | $A_1A_2$ | $A_2A_2$ |
|---|---|---|---|
| $B_1B_1$ | $1-\delta$ | $1-\beta$ | $1-\alpha$ |
| $B_1B_2$ | $1-\gamma$ | $1$ | $1-\gamma$ |
| $B_2B_2$ | $1-\alpha$ | $1-\beta$ | $1-\delta$ |

This model was considered by Bodmer and Felsenstein (1967), followed by more detailed mathematical study by Karlin and Feldman (1969, 1970). We assume that $\alpha$, $\beta$, $\gamma$ and $\delta$ are all positive. Note that model 3 is a special case in which $\alpha = \delta$.

From what we have seen in model 3, we may expect very complex behavior of equilibria and their stability when $\alpha \neq \delta$. This was demonstrated by Karlin and Feldman (1969, 1970). We may generally say that for extremely tight linkage, there always exists a locally stable symmetric equilibrium with $\hat{D} \neq 0$, where by symmetric equilibrium we mean the equilibrium such that $\hat{X}_1 = \hat{X}_4$ and $\hat{X}_2 = \hat{X}_3$. On the other hand, for very loose linkage, a stable equilibrium with $\hat{D} \approx 0$ may exist for any positive set of selection coefficients $\alpha$, $\beta$, $\gamma$ and $\delta$. It may easily be seen that under very tight linkage the situation is analogous to that of a single locus, the two loci forming a "super-gene." This may apply to the case of more than two loci if linkage is very tight and epistasis is strong. On the other hand, under very loose linkage the two loci behave as if independent, each being kept polymorphic by overdominance. However, for some intermediate values of the recombination fraction, and for some sets of selection coefficients, there may be no internal equilibrium or there may exist some unsymmetrical equilibria for which we do not have $\hat{X}_1 = \hat{X}_4$ and $\hat{X}_2 = \hat{X}_3$. The existence of stable unsymmetric equilibria is rather unexpected for a model with such a symmetric viability pattern. Although the biological significance of such unsymmetric equilibria is not very clear, they might have some relevance in considering the fate of newly arisen mutations. An example given by Karlin and Feldman (1969, 1970) is as follows. If $\alpha = 0.03$, $\beta = \gamma = 0.004$, $\delta = 0.005$ and $c = 0.05$ (5% recombination), there are two unsymmetric equilibria which are locally stable, one given by $\hat{X}_1 = 0.8878$, $\hat{X}_2 = \hat{X}_3 = 0.0542$, $\hat{X}_4 = 0.0038$ and another by $\hat{X}_1 = 0.0038$, $\hat{X}_2 = \hat{X}_3 = 0.0542$, $\hat{X}_4 = 0.8878$.

So far we have mainly considered the nature of equilibria using several specific models. Let us now turn to the dynamical aspect of the two-locus problem. Again, we may generally say that for extremely tight linkage, the two loci behave as a single locus, while for very loose linkage, they behave as two independent loci even if there is epistasis, unless it is very strong. However, for an intermediate situation, the path taken by the chromosome frequencies is difficult to determine and no general theories have been obtained. In some cases, there may be several peaks in the adaptive surface, and a slight difference in the initial gene frequencies might lead to an entirely different path.

In natural populations, it must be typical that two segregating loci, chosen at random, are loosely linked and selection coefficients involved are small. For such a situation, we have a fairly general theory which can predict the course of change in chromosome frequencies under random mating. This is the theory of quasi-linkage equilibrium as developed by Kimura (1965a).

According to this theory, if gene frequencies are changing by natural selection under loose linkage and relatively weak epistatic interaction, a state is quickly realized in which the chromosome frequencies change in such a way that the ratio

$$R = \frac{X_1 X_4}{X_2 X_3} \tag{10}$$

remains practically constant.

In a haploid population, if $w_1$, $w_2$, $w_3$ and $w_4$ are respectively the fitnesses of four genotypes $A_1B_1$, $A_1B_2$, $A_2B_1$ and $A_2B_2$, it can be shown that, at the quasi-equilibrium, we have approximately

$$\hat{R} = 1 + \frac{\epsilon}{c}, \tag{11}$$

where $\epsilon = w_1 - w_2 - w_3 + w_4$ is a measure of epistatic interaction in fitness. This formula is valid if $|\epsilon| \ll c$. The state, $R = \hat{R}$, is stable and any deviation of $R$ from $\hat{R}$ will be reduced roughly by fraction $c$ in each generation.

For a diploid population, the situation is more complicated and the corresponding measure of epistatic interaction,

$$\bar{\epsilon} = w_{1\cdot} - w_{2\cdot} - w_{3\cdot} + w_{4\cdot},$$

generally depends on chromosome frequencies so that $\bar{\epsilon}/c$ is not a constant. Nevertheless, quasi-linkage equilibrium may be attained as in the haploid case if gene frequencies are changing slowly under loose linkage and weak selection. In such a quasi-equilibrium $R$ remains nearly constant and it is roughly equal to $1 + \bar{\epsilon}/c$ if $|\bar{\epsilon}| \ll c$.

In the above treatment, departure from random combination of genes between two loci is measured by $R$, the ratio between the frequencies of coupling and repulsion double heterozygotes in the diploid phase.

It is important to note, for a real understanding of the theory, that $R$ is intrinsically different from $D(= X_1X_4 - X_2X_3)$. Although $R = 1$ is equivalent to $D = 0$, constant $R$ is by no means equivalent to constant $D$ (contrary to the assertion by Li 1967, p. 467). Note also that a small $D$ is not necessarily equivalent to a small $|R - 1|$. For example, as demonstrated by Kimura (1965a), in some cases, $D$ tends to zero as $R$ tends to infinity. Generally speaking, $R$ is more sensitive to linkage disequilibrium than $D$ and less dependent on gene frequencies at an individual locus.

The existence of quasi-linkage equilibrium was recently demonstrated more convincingly for a haploid model by Crow and Kimura (1970, p. 198), who could set an exact bounds to $R$ in the course of change in chromosome frequencies. Also, Wright (1967) developed an iterative procedure by which an accurate numerical value of $\hat{R}$ can be obtained for a given set of chromosome frequencies. He then considered the quasi-equilibrium surface of mean selective values. If the system of chromosome frequencies is not on this surface, it will rapidly approach a nearby place on the surface. Then it moves on the surface toward the nearest peak. This means that Wright's concept of adaptive surface is valid in this extended form. Also, it can be shown (Kimura 1965a, Crow and Kimura 1970) that, under quasi-linkage equilibrium, Fisher's fundamental theorem of natural selection holds, so that

$$\Delta\bar{w} = \frac{V_g}{\bar{w}}, \tag{12}$$

where $V_g$ is the additive genetic variance in fitness.

This may be rather unexpected since it is known (cf. Cockerham 1954, Kempthorne 1957) that under random combination of genes, the covariance between parent and offspring is

$$\text{Cov}\,(P, O) = \tfrac{1}{2}V_G + \tfrac{1}{4}V_{AA} + \cdots,$$

where $V_G$ is the additive genetic variance and $V_{AA}$ is the component of epistatic variance due to additive × additive gene interaction. A remarkable fact is that under quasi-linkage equilibrium, just enough linkage disequilibrium is generated to cancel out the contribution from epistatic variance so that the rate of change in fitness at any time is indeed equal to the genic variance at that time. Moran (1964), in his paper entitled "The nonexistence of adaptive topographies" gave an example in which $\bar{w}$ decreases steadily from generation to generation until an equilibrium is reached. It turns out that his example corresponds to a situation in which two inbred lines are crossed. Such a cross is likely to have a selective value higher than the peak on the equilibrium surface and hence the mean fitness decreases through recombination until it reaches the surface. This is clearly not the kind of situation for which Wright has developed his concept of adaptive surface.

Wright (1967) has further shown that even if the epistatic force is not so weak in comparison with the recombination fraction, there may be quasi-equilibrium in a broader sense.

We have seen that the concept of quasi-linkage equilibrium is quite useful to study the situation in which gene frequencies are changing under weak epistatic interaction and loose linkage. The concept is also useful, as shown by Kimura (1966), by enabling us to see two-locus polymorphism in a novel way that cannot be attained through the study of specific models. Namely, it can be shown that the equilibrium set of chromosome frequencies $(\hat{X}_1, \hat{X}_2, \hat{X}_3, \hat{X}_4)$ which can be obtained by setting $\Delta X_i = 0$ in equations (1) is a stationary point of the function $\bar{w}$ with two side conditions

$$\sum_{i=1}^{4} X_i = 1 \tag{13}$$

and

$$\frac{X_1X_4}{X_2X_3} = R_0, \tag{14}$$

where $R_0 = (\hat{X}_1\hat{X}_4)/(\hat{X}_2\hat{X}_3)$ is the value of $R$ at the equilibrium. This can be shown as follows. The stationary point of

$$\bar{w} = \sum_i \sum_j w_{ij}X_iX_j = \sum_{i=1}^{4} w_{i\cdot}X_i$$

with two side conditions (13) and (14) is equivalent to the stationary point of

$$\phi = \sum_{ij} w_{ij}X_iX_j - 2\mu\left(\sum_i X_i - 1\right)$$

$$- 2\lambda\ (\log \frac{X_1X_4}{X_2X_3} - \log R_0), \tag{15}$$

in which no restrictions are imposed among the four chromosome frequencies $X_1$, $X_2$, $X_3$ and $X_4$ ($-2\mu$ and $-2\lambda$ are Lagrange multipliers). Thus, differentiating $\phi$ with respect to $X_i$'s and setting each to zero, we get

$$\begin{aligned}
\partial\phi/\partial X_1 &= 2w_{1\cdot} - 2\mu - (2\lambda/X_1) = 0\\
\partial\phi/\partial X_2 &= 2w_{2\cdot} - 2\mu + (2\lambda/X_2) = 0\\
\partial\phi/\partial X_3 &= 2w_{3\cdot} - 2\mu + (2\lambda/X_3) = 0\\
\partial\phi/\partial X_4 &= 2w_{4\cdot} - 2\mu - (2\lambda/X_4) = 0.
\end{aligned} \tag{16}$$

Multiplying the $i$th equation by $X_i$ and adding them ($i = 1, \ldots, 4$), we find

$$\mu = \bar{w}. \tag{17}$$

Also, subtracting the sum of the second and the third equations from the sum of the first and the fourth equations, we find

$$\lambda = \frac{w_1. - w_2. - w_3. + w_4.}{\sum_{i=1}^{4} (1/X_i)}. \tag{18}$$

Therefore, we have, from (16) and (17),

$$X_1(w_1. - \bar{w}) = - X_2(w_2. - \bar{w}) = - X_3(w_3. - \bar{w})$$

$$= X_4(w_4. - \bar{w}) = \lambda, \tag{19}$$

where $\lambda$ is given by (18). This set of equations together with $(X_1X_4)/(X_2X_3) = R_0$ is equivalent to $\Delta X_1 = \Delta X_2 = \Delta X_3 = \Delta X_4 = 0$ in (1).

Finally, we shall briefly discuss the evolutionary adjustment of linkage intensity. Fisher (1930b) pointed out long ago that epistatic interaction always tends to diminish recombination, while for the spread of advantageous mutations increased recombination may be favored. It is possible that linkage intensity is adjusted by these opposing factors. Recently Nei (1967) has shown that chromosome variants as well as recombination-reducing genes have selective advantage only when epistatic interaction in fitness between loci is involved. He expects that the modification of linkage intensity through such selection always occurs in the direction of decreased recombination. By surveying the recombination fraction for various organisms ranging from viruses to man, he has shown (Nei 1968b) that the recombination fraction is much smaller per unit length of DNA in higher organisms than in lower ones. This appears to be a result of selection for tighter linkage in the process of evolution since genes interact at the various levels of gene function and there are ample opportunities for such selection.

CHAPTER SEVEN

# Linkage Disequilibrium and Associative Overdominance in a Finite Population

In panmictic populations, two factors are mainly responsible for linkage disequilibrium or non-random association of genes between different loci. One is epistatic interaction in fitness, as we have discussed in the previous chapter. Another is random frequency drift caused by sampling of gametes in a finite population, as we shall consider in the present chapter.

Let us assume that a pair of alleles $A_1$ and $A_2$ are segregating with frequencies $p$ and $1-p$ at the first locus, and alleles $B_1$ and $B_2$ with frequencies $q$ and $1-q$ at the second locus.

As an index to measure linkage disequilibrium, we again use $D = X_1X_4 - X_2X_3$, where $X_1$, $X_2$, $X_3$ and $X_4$ are frequencies of four chromosome types $A_1B_1$, $A_1B_2$, $A_2B_1$ and $A_2B_2$ in the population. We shall denote by $N_e$ the "variance" effective size of the population.

As we have seen in the previous chapter, under random mating and additive fitnesses between loci, $D$ approaches zero in an infinitely large population. In a finite population, however, the situation is more complicated, for $D$ fluctuates from generation to generation, even if its long term average is zero. If the effective size of the population is small, such fluctuations may become important in determining the genetic structure of the population.

Let us first consider the mean or the expected value of $D$. Under random mating and no selection, it can be shown that if $c$ is the recombination fraction between the two loci,

$$E(D_t) = D_0 \exp\left\{-\frac{(2N_e c + 1)t}{2N_e}\right\}, \qquad (1)$$

where $D_t$ is the amount of linkage disequilibrium at the $t$th generation and $D_0$ is the initial ($t = 0$) linkage disequilibrium. Namely, the expected value of $D$ decreases at the rate of $c + (1/2N_e)$ per generation until it becomes zero. This may easily be understood by the following consideration. It is well known that $D$ decreases at the rate $c$ per generation in an infinite population, i.e. $D_t = (1 - c)D_{t-1}$. Also, we know that, in a finite population, the average frequency of heterozygotes, or the expectation of any particular heterozygote decreases by a fraction $1/2N$ each generation. Therefore, treating chromosomes as units (i.e. ignoring crossing over), $X_1X_4$ and $X_2X_3$ are both decreasing at this rate; therefore $D_t = (1 - 1/2N)D_{t-1}$. Putting together the effects of recombination and finite size

$$\begin{aligned}\bar{D}_t &= (1 - c)\left(1 - \frac{1}{2N}\right)\bar{D}_{t-1} \\ &\approx \left(1 - \frac{2Nc + 1}{2N}\right)\bar{D}_{t-1}\end{aligned}$$

which is essentially equivalent to (1).

Evaluation of $E(D_t^2)$ or the variance of $D$ is much more difficult, and the explicit expression was first obtained by Hill and Robertson (1968) only for $c = 0$. An essential part of their theory is that if we consider the three quantities,

$$\begin{aligned}X &= E\{pq(1 - p)(1 - q)\} \\ Y &= E\{D(1 - 2p)(1 - 2q)\} \\ Z &= \mathrm{E}\{D^2\},\end{aligned}$$

then the change of the set of three quantities from one generation to the next is given by the equation

$$\begin{bmatrix} X' \\ Y' \\ Z' \end{bmatrix} = \begin{bmatrix} (1-\Lambda)^2 & \Lambda(1-\Lambda)^2(1-c) & 2\Lambda^2(1-\Lambda)(1-c)^2 \\ 0 & (1-\Lambda)(1-2\Lambda)^2(1-c) & 4\Lambda(1-\Lambda)(1-2\Lambda)(1-c)^2 \\ \Lambda(1-\Lambda) & \Lambda(1-\Lambda)^2(1-c) & (1-\Lambda)[\Lambda^2+(1-\Lambda)^2](1-c)^2 \end{bmatrix} \begin{bmatrix} X \\ Y \\ Z \end{bmatrix} \quad (2)$$

where $\Lambda = 1/(2N_e)$ and $X'$, $Y'$, $Z'$ are the values of $X$, $Y$, $Z$ in the next generation.

Hill and Robertson also carried out extensive Monte Carlo experiments and showed that the average value of the squared correlation, or more accurately, mean square contingency of gene frequencies, i.e.

$$r^2 = \frac{D^2}{p(1-p)q(1-q)} \quad (3)$$

approaches a limiting value among lines segregating at both loci, and this value is almost entirely determined by $N_e c$ even with overdominance. In addition, they found the important result that $E(r^2)$ approaches $1/(4N_e c)$ as $N_e c$ becomes large.

Ohta and Kimura (1969a, b) developed a theory based on diffusion models which enables us to treat the problem of linkage disequilibrium in a finite population. They used the Kolmogorov backward equation with respect to the probability distribution of chromosome frequencies. In a simple case of no selection and no mutation they obtained explicit expressions for $X$, $Y$ and $Z$ in an arbitrary generation ($t$) as functions of those values at the initial generation ($t = 0$). Their main results may be summarized as follows. Assuming that $X$, $Y$ and $Z$ can each be expressed as a linear combination of terms of the form

$$e^{\lambda t/N_e},$$

it can be shown that $\lambda$ satisfies the cubic equation

$$\lambda^3 + (5 + 3R)\lambda^2 + \left(\frac{27 + 38R + 8R^2}{4}\right)\lambda + \frac{9 + 26R + 8R^2}{4} = 0, \quad (4)$$

where $R = N_e c$. This equation has three negative roots $\lambda_1$, $\lambda_2$ and $\lambda_3$. Of these three roots, $\lambda_1$, having the smallest absolute value, when divided by $N_e$, gives the asymptotic rate of decrease of $X$, $Y$ and $Z$ as $t$ gets large, so that, for example

$$Z_t = E(D_t^2) \propto e^{\lambda_1 t/N_e}. \quad (5)$$

The value of $\lambda_1$ depends on $N_e c$, and it changes from $-0.5$ to $-1.0$ as $R = N_e c$ changes from 0 to $\infty$. Table 7.1 shows the relation between $|\lambda_1|$ and $N_e c$ for $N_e c$ up to 4.0.

TABLE 7.1. Relation between $N_e c$ (product of the effective population number and the recombination fraction) and $|\lambda_1|$ (the absolute value of the eigenvalue $\lambda_1$). The quantity $|\lambda_1|/N_e$ gives the asymptotic rate of decrease of simultaneous heterozygosity at both loci.

| $N_e c$ | 0.0 | 0.2 | 0.4 | 0.6 | 0.8 | 1.0 | 2.0 | 4.0 |
|---|---|---|---|---|---|---|---|---|
| $\|\lambda_1\|$ | 0.500 | 0.668 | 0.776 | 0.843 | 0.885 | 0.912 | 0.967 | 0.989 |

In considering the behavior of two segregating loci under random drift, the quantity $|\lambda_1|/N_e$ is important since it gives the asymptotic rate of decrease of simultaneous heterozygosity, as well as that of simultaneous segregation at two loci. The problem was also studied by Karlin and McGregor (1968) using the Markov chain method. An explicit expression for $\lambda_1$ was given by Ohta and Kimura (1969a):

$$\lambda_1 = \frac{1}{3}(19 + 6R + 12R^2)^{1/2} \cos\frac{\theta}{3} - R - \frac{5}{3}, \quad (6)$$

where

$$\theta = \cos^{-1}\left\{-\frac{28 + 63R - 90R^2}{(19 + 6R + 12R^2)^{3/2}}\right\},$$

and $R = N_e c$.

As a measure of linkage disequilibrium, $D$ has a shortcoming in that it depends heavily on gene frequencies; so also does $D^2$. Therefore, $r^2$ used by Hill and Robertson is more appropriate than $D^2$ to represent the amount of linkage disequilibrium due to random drift. On the other hand, analytical treatment of this quantity is much more difficult than $D^2$. Therefore, we have used the very similar quantity

$$\sigma_d^2 = \frac{Z}{X} = \frac{E(D^2)}{E\{pq(1-p)(1-q)\}}, \tag{7}$$

which is a standardized or normalized value of $E(D^2)$. The square root of this quantity, i.e. $\sigma_d$, has been termed the standard linkage deviation. Monte Carlo experiments show that $\sigma_d^2$ is usually not very different numerically from $E(r^2)$. So for most purposes, we may equate these two quantities as a sufficiently accurate approximation.

At the state of steady decay, we can show that if $R = N_e c \gg 1$ so that $\lambda_1$ is near to $-1$:

$$\sigma_d^2 \approx \frac{1}{4R + 1 - 3/(R + 1.5)}. \tag{8}$$

Actually, this formula may be used for all $R > 1$. For large values of $R$, the simpler formula

$$\sigma_d^2 \approx \frac{1}{4R} \tag{9}$$

suffices.

An interesting fact is that a similar relation between $\sigma_d^2$ and $R$ holds also for a population in which a stationary distribution of chromosome frequencies is reached under recurrent mutation and random sampling of gametes, as we shall now discuss.

Let the mutation rates be as follows:

$$A_1 \underset{v_1}{\overset{u_1}{\rightleftarrows}} A_2 \qquad\qquad B_1 \underset{v_2}{\overset{u_2}{\rightleftarrows}} B_2.$$

Then, it has been shown by Ohta and Kimura (1969b) that

$$\sigma_d^2 = \frac{1}{3 + 4N_e(c+k) - 2/(2.5 + N_e(c+2k))}, \qquad (10)$$

where $k = u_1 + v_1 + u_2 + v_2$. Thus, for a large $4N_ec$, we have again $\sigma_d^2 \approx 1/(4N_ec)$, provided that $c \gg k$.

As long as there is no selection, the behavior of an individual gene is not influenced by genes at other loci. However, once selection is involved, linkage disequilibrium becomes important, for genes become interdependent in fitness.

Before we go into a detailed discussion of this subject, let us consider a simple case in which a neutral locus is linked to an overdominant locus. Suppose that alleles $B_1$ and $B_2$ are selectively neutral, and $A_1$ and $A_2$ are overdominant, such that the heterozygote $A_1A_2$ has a selective advantage $s_1$ and $s_2$ over the two homozygotes, $A_1A_1$ and $A_2A_2$. In the interest of simplicity let us assume that the overdominance is so strong that the frequency of $A_1(p)$ is kept practically constant at $\hat{p} = s_2/(s_1 + s_2)$. Then it can be shown (Ohta and Kimura 1970) that the squared standard linkage deviation between the $A$ and $B$ loci becomes

$$\sigma_d^2 = \frac{1}{1 + 4N_e(c+k) + \dfrac{(1-2\hat{p})^2}{\hat{p}(1-\hat{p})}\,\dfrac{Ne(c+2k)}{1 + N_e(c+2k)}}. \qquad (11)$$

Thus we have again $\sigma_d^2 \approx 1/(4N_ec)$ for a large $4N_ec$. In addition, it can be shown that this approximation holds when both loci are strongly overdominant. We can conclude, therefore, that as long as $N_ec$ is fairly large, the square of the correlation of gene frequencies due to random genetic drift becomes roughly $1/(4N_ec)$ irrespective of whether there is selection or mutation, or whether it is in a steady state or in a transient state with steady decay. Probably, $N_e$ is sufficiently large in most natural populations that this condition ($4N_ec \gg 1$) is satisfied, although this is not necessarily so for experimental populations.

Now, we shall consider the question of how non-random association influences the behavior of individual genes through interdependence in fitness. Using the previous model of overdominant and neutral loci, let us see how apparent selection develops at the neutral locus ($B$).

In this case, the apparent selection at the $B$ locus takes the form of "associative overdominance" (Frydenberg 1963). Let $s_1$ and $s_2$ be the selection coefficients against the two homozygotes at $A$ locus and let $s_1'$ and $s_2'$ be those at $B$ locus. Then we can show (Ohta and Kimura 1970) that

$$s_1' = E(W_{B_1B_2}) - E(W_{B_1B_1}) = E\left\{(-s_2 + (s_1 + s_2)p)\frac{D}{q(1-q)} + (s_1 + s_2)\frac{D^2}{q^2(1-q)}\right\} = (s_1 + s_2)E\left(\frac{D^2}{q^2(1-q)}\right)$$

and

$$s_2' = E(W_{B_1B_2}) - E(W_{B_2B_2}) = E\left\{(s_2 - (s_1 + s_2)p)\frac{D}{q(1-q)} + (s_1 + s_2)\frac{D^2}{q(1-q)^2}\right\} = (s_1 + s_2)E\left(\frac{D^2}{q(1-q)^2}\right), \tag{12}$$

because $E(D) = 0$. With symmetrical overdominance ($s_1 = s_2$), assuming that $q$ takes an intermediate value,

$$s_1' = E(W_{B_1B_2}) - E(W_{B_1B_1}) = 2s\sigma_d^2\frac{p(1-p)}{q}$$

$$s_2' = E(W_{B_1B_2}) - E(W_{B_2B_2}) = 2s\sigma_d^2\frac{p(1-p)}{1-q}.$$

Furthermore, if we assume $p = \hat{p} = 1/2$ and an arbitrary value of $q = 1/2$, we have

$$s_1' = s_2' = s\sigma_d^2. \tag{13}$$

This is a rather small quantity, but with many linked overdominant loci, the cumulative effect may not be negligible.

Such an effect of associative overdominance must be taken into account not only for the neutral loci but also for the overdominant loci themselves. The apparent selective values at a locus, then, are the sum of its own selective values and the effect of neighboring loci.

In order to find out how much associative overdominance will be developed if a neutral locus is linked with a number of truly overdominant loci, let us consider a simple model as shown in Figure 7.1.

In this model, we assume that, on the left of the neutral locus, there are $n_1$ overdominant loci spaced at intervals of one crossover unit (one percent recombination), and similarly $n_2$ overdominant loci on the right. The distance between the neutral and the nearest overdominant locus on either side is also one crossover unit.

Let $s$ be the selection coefficient against either homozygote at each overdominant locus (symmetric overdominance), and suppose that the two alleles $B_1$ and $B_2$ at the neutral locus happen to be about equal in frequency ($q \approx 0.5$).

Then the "apparent" selective advantage of the hetero-

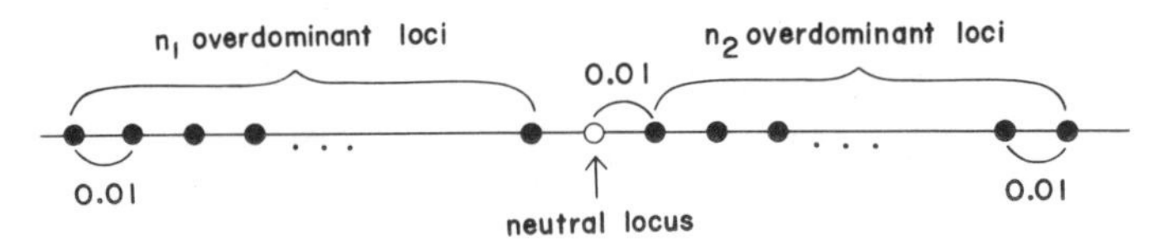


FIGURE 7.1. A model used to investigate the effect of associative overdominance.

zygote over either homozygote at the neutral locus becomes approximately

$$s' = \frac{100s}{4N_e}\{2\gamma + \log n_1 + \log n_2\} \tag{14}$$

where $\gamma = 0.577\ldots$, which is Euler's constant. In deriving this formula, we assume additive overdominance between loci and that $\sigma_d^2$ between the neutral locus and one of the overdominant loci is given by $1/(4N_ec)$, where $c$ is the recombination fraction between them. We also assume that non-random association between overdominant loci is not so strong as to make the whole system organized in a "super-gene."

For example, if $n_1 = n_2 = 50$ and $N_e = 1{,}000$, $s'$ becomes about $0.225s$. If the effective size of the population is 10 times as large, $s'$ is $0.0225s$. In the former case, the apparent overdominance is about one fourth as large as the true overdominance. Also, it is important to note that $N_es'$ remains constant with varying $N_e$.

To check the validity of these approximations for multilocus systems, we used Monte Carlo experiments. The model used was as follows. A total of 23 equally spaced loci are assumed (see Figure 7.1). The recombination fraction $c$ between two adjacent loci varies from 0.0 to 0.03. Forward and reverse mutations between a pair of alleles at each locus occur at an equal rate $u = 0.005$ per generation. The three loci, terminal 1, terminal 2 and the middle locus are chosen as markers, of which terminal 1 and the middle locus are neutral. All the remaining 21 loci are assumed to be overdominant with $s_1 = s_2 = 0.01$. In Figure 7.2 the experimental results are shown by dots and theoretical predictions are given by curves. The latter have been computed by using formulae essentially similar to equation (14). As may be seen from the figure, the agreement between the experimental results and theoretical predictions is satisfactory.

If linkage is much tighter and overdominance at each locus is much stronger, genes on a chromosome may

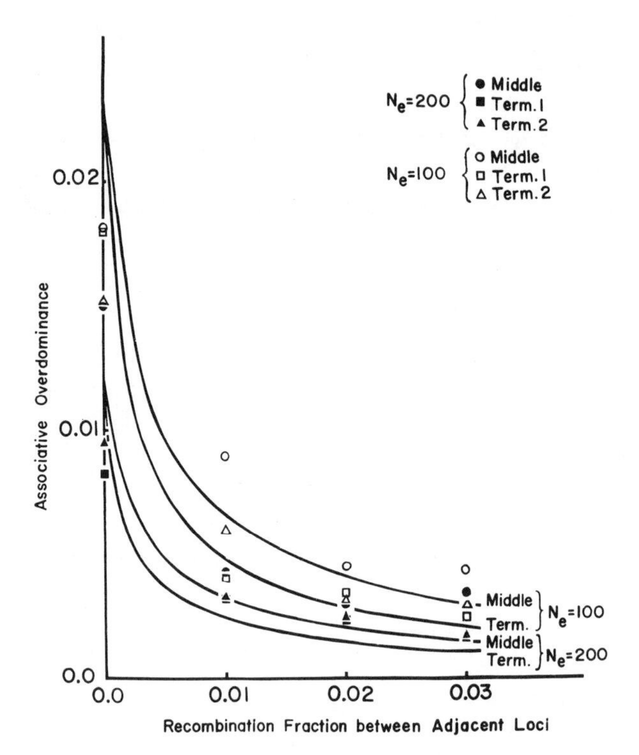


FIGURE 7.2. Associative overdominance observed at the three marker loci. The results of Monte Carlo experiments are given by dots and the theoretical predictions by curves. The marker loci (terminal 1, terminal 2 and middle) are located on the chromosome similar to Figure 7.1 with $n_1 = n_2 = 11$. A total of 23 loci are equally spaced and all the loci except terminal 1 and middle loci are overdominant with $s_1 = s_2 = 0.01$. Population size ($N_e$) is either 100 or 200, and the mutation rates are $u_1 = v_1 = u_2 = v_2 = 0.005$ giving $k = 0.02$. The recombination fraction between adjacent loci (c) changes from 0.0 to 0.03. The experimental values are the averages through 501st to 1,000th generations.

become organized as a "super-gene" and the above treatment will no longer be valid.

Recently, the magnitude of associative overdominance due to linked detrimental mutations has been estimated (Ohta, 1971). Actually, the similar relationship between $s'$ and inbreeding depression holds approximately irrespective of whether it is mutational or segregational. Therefore the effect of linked detrimental mutations is probably more important than linked overdominant alleles as the actual cause of associative overdominance.

The problem of linkage disequilibrium is particularly pertinent in experimental populations. Numerous experiments have been performed for the purpose of estimating fitness values with respect to isozyme alleles or other marker genes. Very often, however, the results merely reflect the effects of a group of surrounding genes, and therefore, the effect of individual alleles on fitness is quite difficult to measure. In experimental populations, the initial linkage disequilibrium may be produced by sampling a relatively small number of chromosomes from a large parental population. Suppose that the linkage disequilibrium in the parental population is negligible and that $n$ chromosomes are sampled. Then, applying formula (2) with $c = 0$, the squared standard linkage deviation created by sampling becomes

$$\sigma_d^2 = \frac{1}{n-1}. \tag{15}$$

Usually, the chromosomes thus sampled are rapidly multiplied from $n$ to $n'$ ($n' \gg n$) in the succeeding generations. In such cases, $\sigma_d^2$ changes from $1/(n-1)$ to about $1/(2n'c)$ if $n'c \gg 1$. A seemingly large selective value which lasts for only a few generations at the beginning of an experiment is probably a by-product of linkage disequilibrium between the marker and some other selected genes.

Finally, we would like to mention another topic which is related to linkage in finite populations. This is the effect of linkage on fixation probability. Hill and Robertson (1966) carried out extensive Monte Carlo experiments to study the effect of linkage on selection limits. With additive gene effects, i.e. with no dominance and no epistasis, the effect of linkage on the fixation probability is negligible if the selection coefficient, $s$, is small. However, if $s$ is large and $N_e s$ is much larger than unity linkage may have a considerable effect in reducing the fixation probability. If an allele at the first locus is strongly favored, the average probability of fixation of a favorable allele at the second locus is reduced when linkage is tight. Hill and Robertson conclude that the reduction is not due to linkage disequilibrium but it is due to reduction in effective population number. This may be understood by considering the following situation. If a favorable allele is eventually fixed at a locus, the effective population number will be reduced for the nearby locus having a smaller effect, for gene frequency change has to be taken place within a lineage containing the allele strongly selected for.

CHAPTER EIGHT

# Breeding Structure of Populations

In mathematical treatments of population genetics it is often assumed that the population is a single panmictic unit, that is to say, mating takes place as if mates were randomly chosen from the entire population. This assumption has the welcome property of greatly simplifying the mathematics. Furthermore, it is usually quite realistic for populations that are not spread over too large an area and Hardy-Weinberg frequencies are often observed in actual data.

On the other hand, the total population forming a species usually cannot be a random mating unit, especially when the species occupies a very large territory. This is because the distance of individual migration is much smaller than the entire distribution range of the species. Wright (1943) used the term "isolation by distance" to denote such a phenomenon. This may lead to local differentiation of gene frequencies, even if the distribution of individuals is continuous.

If, in addition, geographical barriers exist between subpopulations, isolation between them becomes more effective. It is well known that such barriers promote formation of races and new species. In fact, Wagner (1868) long ago emphasized the importance of isolation in evolution.

The first important contribution to the mathematical treatment of subdivided populations in relation to local differentiation of gene frequencies was made by Wright (1940, 1943, 1951). In his treatment, the basic concept is the correlation between uniting gametes, calculated by assigning the value 1 to gene $A$ and value 0 to its allele $a$ (or some other arbitrary value). It is called the inbreeding coefficient, and is usually denoted by $F$ or $f$. Wright devised a set of $F$-coefficients for a subdivided population and discovered an important relation among them: Suppose that the total population ($T$) is divided into a certain number of subpopulations ($S$) or demes. Then the inbreeding coefficient of an individual ($I$) relative to the total population ($T$) is given by

$$F_{IT} = F_{ST} + (1 - F_{ST})F_{IS}, \tag{1}$$

where $F_{IS}$ is the inbreeding coefficient of an individual relative to the subpopulation in which it is contained, and $F_{ST}$ is that of the subpopulation relative to the total population; $F_{ST}$ is the correlation between random gametes drawn from the same subpopulation. In terms of panmictic index $P = 1 - F$, the relationship can be expressed in a more convenient form:

$$P_{IT} = P_{IS}P_{ST}. \tag{2}$$

Since the inbreeding coefficient may be interpreted as the probability that two homologous genes in uniting gametes are descended from a common ancestral gene (Malécot 1948), it is of interest to derive this formula from this probability consideration. As pointed out by Crow and Kimura (1970), the left-hand side of formula (2), i.e. $P_{IT}$ or $1 - F_{IT}$, represents the overall probability that two homologous genes in $I$ are not identical by descent, and this is equal to the product of the two probabilities $P_{IS}$ and $P_{ST}$: The former is the probability that two homologous genes in $I$ are not identical by descent relative to $S$, and the latter is the probability that two randomly chosen homologous genes in $S$ are not identical by descent.

From the above interpretation, it is evident that $F_{IS}$ gives the effect of local inbreeding such as consanguineous mating within $S$, while $F_{ST}$ gives the effect of gene frequency differentiation among subpopulations caused by random genetic drift. $F_{ST}$ is not zero even if mating within each $S$ is random. It is sometimes called the random component of $F_{IT}$, while $F_{IS}$ is called the non-random component.

Although pedigree studies are required in general to compute $F$ statistics for a given population, there is an elegant shortcut that may be applicable to some human populations. It was developed by Crow and Mange (1965) based on the observation that in many human societies the surname is transmitted in a regular pattern such that the frequency of marriages between persons of the same surname (isonymous marriage) may be used to measure inbreeding. Namely, if $A$ is the fraction of isonymous marriages, then $F = A/4$ for a majority of consanguineous marriages. For example, if the marriage is between the first cousins, then it is well known that $F = 1/16$, while $A = 1/4$ since they have the same surname only when they are related through their fathers. Crow and Mange then showed that if $a_i$ and $b_i$ are respectively the proportions of the $i$th surname in males and females,

$$F_{IS} = \frac{(1/4)(A - \Sigma a_i b_i)}{1 - \Sigma a_i b_i} \tag{3}$$

and

$$F_{ST} = (1/4)\Sigma a_i b_i \tag{4}$$

approximately, so that the total inbreeding may be computed from formula (1). They applied this method to the Hutterite population and got $F_{IS} = 0.0052$, $F_{ST} = 0.0445$, $F_{IT} = 0.0495$, suggesting that the total inbreeding effect is almost entirely due to random genetic drift. According to Crow (personal communication), a slight modification of formula (3) to make

$$F_{IS} = \frac{(1/4)(A - \Sigma a_i b_i)}{1 - \Sigma a_i b_i / 4} \tag{3'}$$

will improve the estimations, giving $F_{IS} = 0.00437$, $F_{ST} = 0.0446$, $F_{IT} = 0.0488$. However, the difference is very slight.

Furthermore, the average value of $F$ computed directly by taking account of the known inbreeding up to fourth cousins turned out to be 0.0226. This means that more than half of the total inbreeding effect is attributable to a common ancestry more remote than fourth cousins.

The inbreeding coefficient $F_{ST}$ can of course be calculated directly, if gene frequencies among subpopulations are known. Namely, if the mean and the variance of the frequencies of a particular allele among subgroups are $\bar{p}$ and $\sigma_p{}^2$ respectively, then

$$F_{ST} = \frac{\sigma_p{}^2}{\bar{p}(1 - \bar{p})} \tag{5}$$

(Wright 1943). Cavalli-Sforza, Barrai and Edwards (1964) applied this formula to the distribution of various blood group gene frequencies among villages in the Parma Valley (Italy) and obtained $F_{ST} = 0.0356 \pm 0.006$.

Recently, Wright's formula (5) was extended by Nei (1965) to cover a multiallelic system. He showed that

$$F_{ST} = \frac{-\sigma_{jk}}{\bar{p}_j \bar{p}_k}, \tag{6}$$

where $\sigma_{jk}$ is the covariance between the frequencies of the $i$th and $j$th alleles, and $\bar{p}_j$ and $\bar{p}_k$ are their mean frequencies. Nei and Imaizumi (1966a) obtained six estimates by applying this formula, as well as formula (5), to ABO blood group gene frequencies among 45 prefectures in Japan. These estimates turned out to be remarkably similar (about

0.0007), suggesting that differentiation of the gene frequencies occurred largely at random.

In order to investigate the effect of population subdivision on the genetic composition of a Mendelian population, it is necessary to have models of population structure. For this purpose, Wright (1940, 1951) proposed two contrasting models. One is the "island model" and the other is the model of continuous distribution in which "isolation by distance" plays the essential role.

In the first model, the total population (species) consists of an array of subgroups (island population) of effective size $N_e$, within each of which mating takes place at random, and individuals are exchanged with the species as a whole at the rate $m$ per generation. Note that the immigrants are drawn at random from the whole population and there is no tendency for them to come from nearby subgroups. Such a situation may be met approximately by a group of oceanic islands and therefore the model is called the island model.

The amount of random differentiation of gene frequencies among subgroups is given by the variance $\sigma_p^2$ and it was shown by Wright (1931 and later) that

$$\sigma_p^2 = \frac{\bar{p}(1-\bar{p})}{4N_e m + 1}, \tag{7}$$

where $p$ is the frequency of a particular allele in a subgroup. Thus, from (5),

$$F = \frac{1}{4N_e m + 1}. \tag{8}$$

A more exact treatment (Wright 1943) gives

$$\sigma_p^2 = \frac{\bar{p}(1-\bar{p})}{2N_e - (2N_e - 1)(1-m)^2}, \tag{9}$$

but the difference is negligible as long as $m$ is small. Figure 8.1 shows the probability distribution of gene frequencies among subpopulations given by Wright (1940) for various values of $F$, assuming $\bar{p} = 0.5$. The amount of differentiation is appreciable even with such a small value of $F$ as 0.05. Random fixation of alleles starts to occur at $F = 0.33$.

In the second model considered by Wright, a population is distributed uniformly over a large territory, but the parents of any given individual are drawn at random from a small surrounding region. The most important parameter involved is the "size of neighborhood," i.e. the effective population size of such a surrounding region. According to Wright (1951), the neighborhood size $N_N$ is approximately equal to the effective number of individuals

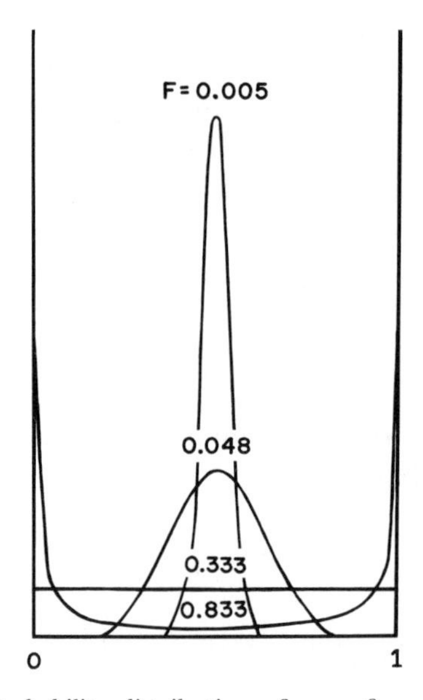


FIGURE 8.1. Probability distribution of gene frequencies among subgroups for various levels of the inbreeding coefficient $F$. Abscissa: gene frequency; ordinate: probability density. (From Wright 1940.)

within a circle of radius twice the standard deviation of the length of migration in one direction per generation. This appears to be valid largely irrespective of the form of the migration distribution. Using this model, he studied the amount of local differentiation in terms of $F_{XT}$, i.e. the inbreeding coefficient of subgroups of various sizes ($N_X$) within the total population ($N_T$) for various values of $N_N$.

It turned out from his analysis that random local differentiation is slight for an area continuum, unless the size of neighborhood is small. Only when $N_N$ is of the order of 20 or less is marked differentiation expected; with $N_N = 200$, the amount of differentiation is fairly small. If $N_N = 1{,}000$ or more, it is so slight that the situation is essentially equivalent to that of universal panmixia. In Chapter 9, we shall discuss this topic again based on the work of Maruyama (1970a, b, c, 1971a).

On the other hand, if the distribution range is essentially one dimensional, marked differentiation may occur even with a relatively large neighborhood such as $N_N = 10^4$.

The problem of local differentiation may also be attacked by studying the pattern of change in correlation with distance as first done by Malécot (1948). With local differentiation, it is expected that individuals living nearby tend to be more alike genetically than those living far apart. His important work on this subject has been recently reviewed and refined (Malécot 1967).

The "coefficient of kinship" $\phi(y)$ between two individuals that are a distance $y$ apart plays a central role in his analysis, where the coefficient of kinship is the probability that two homologous genes, randomly extracted, one from each of the two individuals, are identical by descent.

Consider a habitat where individuals are distributed continuously with equal density $\delta$. Let $\ell(x)$ be the probability density that the distance from the birthplace of the child to that of its parent is $x$, where $x$ is a vector, and in the case of a two dimensional habitat, it has coordinates $x_1$ and $x_2$. This distance is called the "parental distance." Then, the coefficient of kinship in the $n$th generation is related to that of the previous generation by

$$\phi_n(y) = (1-u)^2\{\int \phi_{n-1}(y+z-x)\ell(x)\ell(z)\,dzdx + \frac{(1+f_0)/2 - \phi_{n-1}(0)}{\delta}\int \ell(x)\ell(x-y)\,dx\}, \quad (10)$$

where $u$ is the mutation rate and $f_0$ is the inbreeding coefficient of a parent.

This equation may be derived from the following consideration. Let $I$ and $J$ be the two individuals in the $n$th generation, and let $y$ be the distance of $J$ from $I$. Also, let $P_I$ and $P_J$ be the parents of $I$ and $J$. If we denote by $x$ the vectorial distance of $P_I$ from $I$, and similarly, by $z$ the distance of $P_J$ from $J$, then the vectorial distance of $P_J$ from $P_I$ is $y+z-x$ (see Figure 8.2). Since the probability corresponding to the two parental migrations is $\ell(x)\ell(z)$ and since the coefficient of kinship between $P_I$ and $P_J$ is $\phi_{n-1}(y+z-x)$, the contribution of $P_I$ and $P_J$ to the kinship of $I$ and $J$ is $\phi_{n-1}(y+z-x)\ell(x)\ell(z)$.

The first integral in the braces of the right-hand side of equation (10) is the sum total of such contributions for all possible values of $x$ and $z$. This takes care of the cases

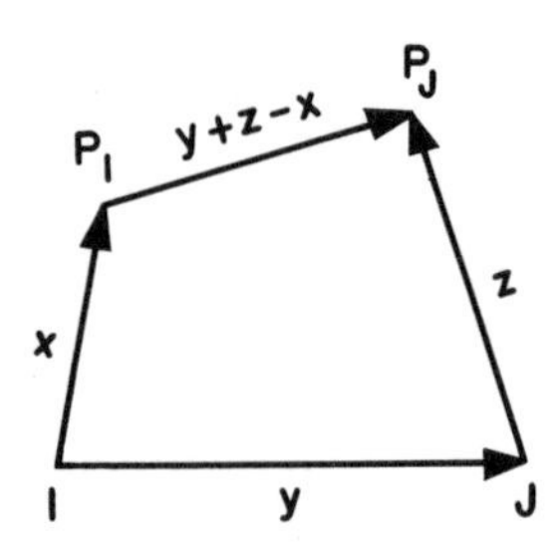


FIGURE 8.2. Distances between four individuals $I$, $J$, $P_I$ and $P_J$. (Modified from Malécot 1967.)

in which the two parents $P_I$ and $P_J$ are distinct. The other term in the braces accompanying the second integral is the correction to be made to include the cases in which $P_I$ and $P_J$ represent the same individual. The probability that they are born in the same elementary area d$x$, which is distance $x$ from $I$, is $\ell(x)\ell(z)\,dx = \ell(x)\,dx\ell(x-y)\,dx$, since $y+z-x=0$ when $P_I$ and $P_J$ are the same. Since there are $\delta\,dx$ individuals in this area, the probability that $P_I$ and $P_J$ are the same and located at distance $x$ from $I$ is $\ell(x)$ $\ell(x-y)\,dx/\delta$.

Since two gametes randomly extracted from the same individual contain either the same gene with probability $1/2$ (in which case the coefficient of kinship is 1), or two different but homologous genes with probability $1/2$ (in which case the coefficient of kinship is equal to the inbreeding coefficient $f_0$), we have to substitute $(1+f_0)/2$ for $\phi_{n-1}(0)$ in the cases in which $P_I$ and $P_J$ represent the same individual. Thus the sum of $[(1+f_0)/2-\phi_{n-1}(0)]\ell(x)$ $\ell(x-y)\,dx/\delta$ for all possible values of $x$ gives the required correction as given by the second term in the braces. In addition, the factor $(1-u)^2$ has to be multiplied through these terms, because the above arguments are valid only when neither gamete carries a mutated gene.

Essentially the same equation as (10) was given earlier by Malécot (1955). Assuming the normal distribution of the distance of parental migration, he showed that, at equilibrium ($\phi_n(y)=\phi_{n-1}(y)\equiv\phi(y)$) and under local random mating ($\phi(0)=f_0$), the inbreeding coefficient is

$$f_0 = \frac{1}{1+4\sigma\delta\sqrt{2u}} \tag{11}$$

for the one-dimensional habitat, and

$$f_0 = \frac{1}{1+8\pi\delta\sigma^2(-1/\log_e 2u)} \tag{12}$$

for the two dimensional habitat, where $\sigma^2$ is the variance of parental migration distance along each axis. Thus the inbreeding coefficient depends strongly on the number of dimensions.

On the other hand, Malécot arrived at the conclusion that the coefficient of kinship $\phi(x)$ is approximately the same for any number of dimensions so that

$$\phi(x) \approx e^{-x\sqrt{2u}/\sigma}. \tag{13}$$

However, he later showed (Malécot 1959) that for the two dimensional case,

$$\phi(x) \propto \frac{e^{-x\sqrt{2u}/\sigma}}{\sqrt{x}} \tag{14}$$

is more appropriate, so there is an effect of dimension.

Malécot (1967) also introduced what he called the "$K$ distribution" to represent the probability distribution of the absolute distance $r$ for the case of two dimensional isotropic migration and showed that essentially the same formulae as (12) and (14) hold also for this distribution. His $K$ distribution,

$$\frac{K_\beta(rh)(hr)^{\beta+1}h}{2^\beta\Gamma(\beta+1)}, \qquad \beta > -1, \tag{15}$$

in which $K_\beta$ denotes the modified Bessel function of the second kind, has a very convenient property that every convolution of $K$ distributions with the same parameter $h$ is also a $K$ distribution, since its characteristic function (Hankel transform) is $\{h^2/(h^2+v^2)\}^{\beta+1}$. It is probable that this distribution will prove to be very useful for the theoretical study of population structure.

Let us now turn to another model of population structure which was termed "the stepping stone model" (Kimura 1953). This is somewhat intermediate in character between Wright's island model and the model of a continuum. In this model, the total population is subdivided into discrete colonies as in the island model, but individuals are exchanged only between adjacent colonies. The genetical properties of this model were clarified by Kimura and Weiss (1964), followed by a more complete mathematical analysis by Weiss and Kimura (1965). They succeeded in obtaining the exact solutions with respect to the variances and covariances at equilibrium. Figure 8.3 illustrates a one dimensional stepping stone model with an infinite number of colonies arranged single file. It is assumed that each colony has effective size $N_e$ and exchanges individuals only with its two neighboring colonies in each generation at the rate $m_1$, such that $m_1/2$ is the proportion of individuals exchanged between a pair of adjacent colonies per generation.

Consider a pair of alleles $A$ and $a$ and let $p_i$ be the frequency of $A$ in the $i$th colony. It is convenient to assume that $p_i$ stands for the frequency at fertilization and that mutation, migration and random sampling of gametes occur in this order to produce the next generation.

Let $u$ be the mutation rate from $A$ to $a$, and let $v$ be that in the reverse direction. Then, by mutation, the frequency of $A$ changes from $p_i$ to

$$p_i(1-u) + v(1-p_i) \tag{16}$$

or

$$p_i + \bar{m}_\infty(\bar{p} - p_i), \tag{16'}$$

where $\bar{m}_\infty = u + v$ and $\bar{p} = v/(u+v)$. Expression (16′) is more convenient than (16), since by writing it this way, not

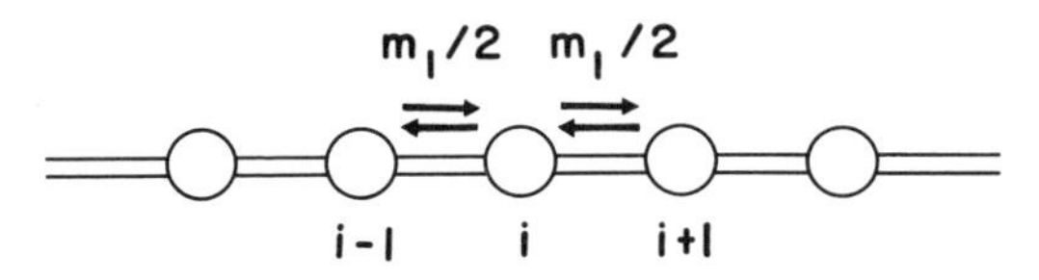


FIGURE 8.3. One dimensional stepping stone model.

only mutation but also long range migration (as in the island model) may be incorporated in the coefficient $\bar{m}_\infty$.

Next, by exchange of individuals between adjacent colonies (i.e. by short range migration characteristic of the stepping stone model), the gene frequency changes further to

$$P_i = (1 - m_1)\{p_i + \bar{m}_\infty(\bar{p} - p_i)\} + \frac{m_1}{2}\{p_{i-1} + \bar{m}(\bar{p} - p_{i-1})\}$$

$$+ \frac{m_1}{2}\{p_{i+1} + \bar{m}(\bar{p} - p_{i+1})\}$$

$$= (1 - m_1)(1 - \bar{m}_\infty)p_i + \tfrac{1}{2}m_1(1 - \bar{m}_\infty)(p_{i-1} + p_{i+1})$$

$$+ \bar{m}_\infty \bar{p}. \tag{17}$$

Finally, by random sampling of gametes, the frequency in the next generation becomes

$$p_i' = P_i + \xi_i, \tag{18}$$

where $\xi_i$ is the amount of change in the frequency of $A$ due to random sampling such that $\xi_i$ has mean 0 and variance $P_i(1 - P_i)/(2N_e)$.

From equation (18) connecting the gene frequency in the present generation with that of the next generation, we can derive an infinite set of equations containing variances and covariances of gene frequencies among colonies. It can then be shown that at equilibrium, in which these quantities do not change with time, the variance of gene frequencies among colonies becomes

$$\sigma_p^2 = \frac{\bar{p}(1 - \bar{p})}{2N_e - (2N_e - 1)\{1 - 2R_1R_1/(R_1 + R_2)\}}, \tag{19}$$

where $R_1 = [(1 + \alpha)^2 - (2\beta)^2]^{1/2}$, $R_2 = [(1 - \alpha)^2 - (2\beta)^2]^{1/2}$ in which $\alpha = (1 - m_1)(1 - \bar{m}_\infty)$ and $\beta = m_1(1 - \bar{m}_\infty)/2$. In the special case of $m_1 = 0$, we have $\alpha = 1 - \bar{m}_\infty$ and $\beta = 0$

so that (19) reduces to Wright's formula (9), except that letter $m$ rather than $\bar{m}_\infty$ is used in the latter to represent the rate of long range migration.

Such an agreement is expected because Wright's island model is a special case of the stepping stone model in which there is no short range migration.

On the other hand, if the rate of short range migration $m_1$ is much higher than that of the long range migration $\bar{m}_\infty$, formula (19) reduces approximately to

$$\sigma_p^2 = \frac{\bar{p}(1-\bar{p})}{1 + 4N_e\sqrt{2m_1\bar{m}_\infty}}, \tag{20}$$

where we assume that $\bar{m}_\infty \ll m_1 \ll 1$.

If we note that, in the stepping stone model, the variance of the migration distance per generation in one direction is $\sigma^2 = m_1$ (one step either to the right or to the left each with frequency $m_1/2$) and the density is $\delta = N_e$ ($N_e$ breeding individuals per unit length), formula (20) leads to

$$f = \frac{\sigma_p^2}{\bar{p}(1-\bar{p})} = \frac{1}{1 + 4\delta\sigma\sqrt{2\bar{m}_\infty}}. \tag{21}$$

This is equivalent to Malécot's formula (11), because in his treatment only mutation from $A$ to other alleles is considered so that $\bar{m}_\infty = u$.

Also, it can be shown for the stepping stone model that at equilibrium the correlation of gene frequencies between two colonies $j$ steps apart is

$$r_j = \frac{\dfrac{1}{2\pi}\displaystyle\int_0^{2\pi} \frac{\cos j\theta \, d\theta}{1 - H^2(\cos\theta)}}{\dfrac{1}{2\pi}\displaystyle\int_0^{2\pi} \frac{d\theta}{1 - H^2(\cos\theta)}}, \tag{22}$$

where $H(\cos\theta) = \alpha + 2\beta\cos\theta = (1 - m_1)(1 - \bar{m}_\infty) + m_1(1 - \bar{m}_\infty)\cos\theta$. Although the integrals can readily be evaluated in terms of elementary functions, we shall not take trouble to do so here except to note that for the case corresponding to the island model ($m_1 = 0$) we have $r_j = 0$ ($j \neq 0$) as expected, while if $m_1 \gg \bar{m}_\infty$, formula (22) reduces approximately to

$$r_j \approx \exp\left\{-j\sqrt{\frac{2\bar{m}_\infty}{m_i}}\right\}. \tag{23}$$

This is equivalent to Malécot's formula (13) as may be seen by substituting $\sigma^2$ for $m_1$ and $u$ for $\bar{m}_\infty$ in the formula.

The above treatments can be extended to cover two and higher dimensional cases. Figure 8.4 illustrates the two dimensional case in which the entire population consists of a rectangular array of colonies extending to infinity in horizontal and vertical directions. In this figure, $m_{1A}$ stands for the rate of short range migration in the horizontal direction and $m_{1B}$ is that in the vertical direction. Thus each colony exchanges individuals with its four neighboring colonies at the rate $m_1 = m_{1A} + m_{1B}$ per generation.

At equilibrium, the variance and covariances of gene

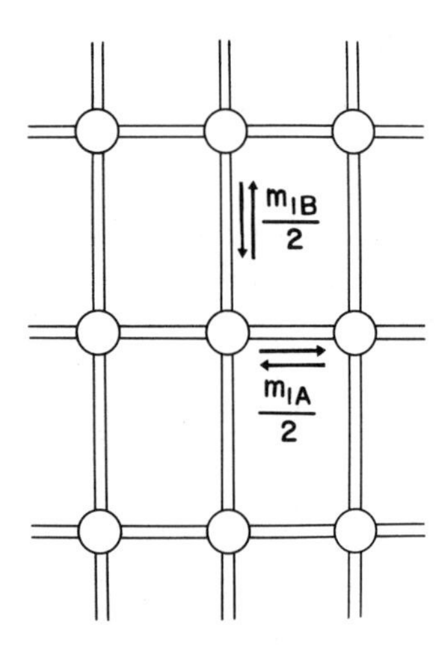


FIGURE 8.4. Two dimensional stepping stone model.

frequencies between colonies can be worked out as in the previous case. Of particular interest is the correlation coefficient of gene frequencies between colonies. It can be shown that if $\bar{m}_\infty$ is much smaller than $m_{1A}$ and $m_{1B}$, which are themselves smaller than unity, the correlation coefficient $r(k_1, k_2)$ between colonies which are $k_1$ steps apart in the horizontal and $k_2$ steps apart in the vertical direction becomes approximately

$$r(k_1, k_2) = \frac{C_0}{2\pi\sqrt{m_{1A}m_{1B}}} K_0(\sqrt{2\bar{m}_\infty}\,\zeta), \tag{24}$$

where $K_0(\cdot)$ stands for the modified Bessel function of zero order, $C_0$ is a constant, and

$$\zeta = \sqrt{\frac{k_1^2}{m_{1A}} + \frac{k_2^2}{m_{1B}}}.$$

The above formula (24) is valid when $\zeta$ is large. If in addition the rates of migration in both directions are equal so that $m_{1A} = m_{1B} = m_1/2$, the formula may be replaced by

$$r(\rho) \propto \frac{e^{-\rho\sqrt{4\bar{m}_\infty/m_1}}}{\sqrt{\rho}}, \tag{25}$$

where $\rho = \sqrt{k_1^2 + k_2^2}$ is the distance between two colonies. This formula is valid when $\rho$ is large, and it is equivalent to Malécot's result given as formula (14). Note that in his formula, $\sigma^2$ is the variance of parental migration in one direction and therefore corresponds to $m_1/2$ in the two dimensional case.

Similarly, we can consider a three dimensional stepping stone model with a cubic array of colonies extending to infinity in all directions. Let $m_{1A}$, $m_{1B}$ and $m_{1C}$ be respectively the rate of short range migration along $X$, $Y$ and $Z$ axes. Each colony exchanges individuals with six adjacent colonies at the rate $m_1 = m_{1A} + m_{1B} + m_{1C}$. Although the exact solution has been obtained for this model, it contains triple integrals that can only be evaluated in general by numerical integration. However, a simple formula is available if the rates of migration are equal in all directions ($m_{1A} = m_{1B} = m_{1C} = m_1/3$) and if $\bar{m}_\infty$ is much smaller than $m_{1A}$. In such a case, the correlation of gene frequencies between two colonies which are distance $\rho$ apart is approximately

$$r(\rho) \approx \frac{e^{-\rho\sqrt{6\bar{m}_\infty/m_1}}}{\pi\rho}. \tag{26}$$

The above treatments of the stepping stone model show that the correlation of gene frequencies between colonies falls off more quickly in the three dimensional case than in the two dimensional case, which in turn falls off more quickly than the one dimensional case. This shows that the tendency toward random local differentiation is strongest in one dimension, but decreases rapidly as the number of dimensions increases. Figure 8.5 illustrates the decrease of genetic correlation with distance assuming $\bar{m}_\infty = 4 \times 10^{-5}$ and $m_1 = 0.1$, namely $m_1 = 0.1$ for one dimension, $m_{1A} = m_{1B} = 0.05$ for two dimensions and $m_{1A} = m_{1B} = m_{1C} = 0.333$ for three dimensions.

A new mathematical treatment of the stepping stone model has recently been developed by Maruyama (1969, 1970c, 1971a), who extended the results to a finite number of colonies and non-symmetric migration rates. We shall discuss some of his latest results in Chapter 9 in relation to patterns of neutral polymorphism in geographically structured populations.

Man is probably the best material for studying the breeding structure of populations, since genealogical records of individuals and demographic data in any district are more complete and easily available than for any other animal. Using parish records in the Parma Valley

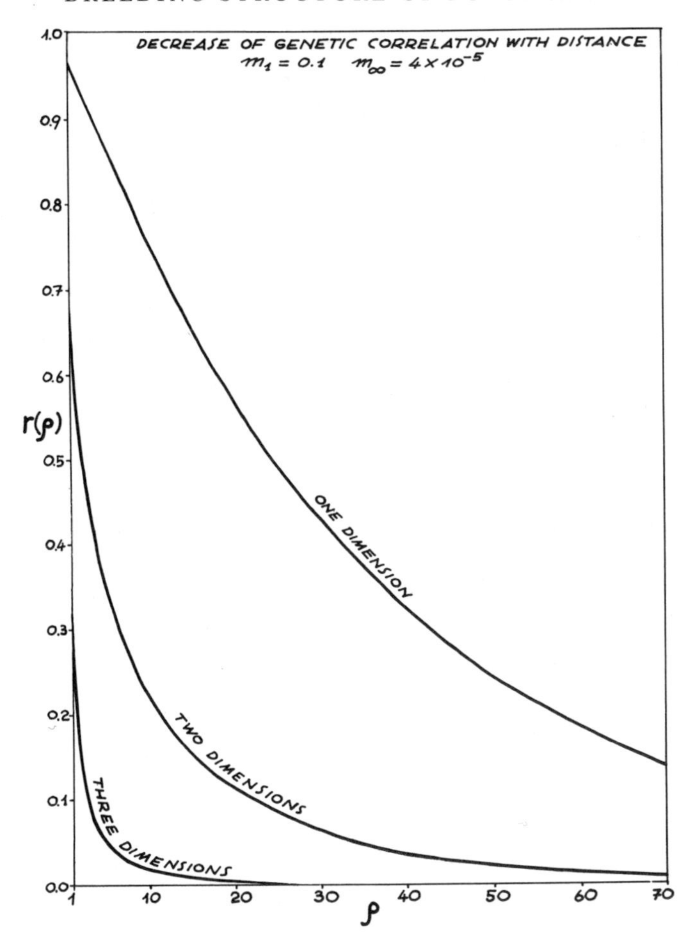


FIGURE 8.5. Decrease of genetic correlation $r(\rho)$ between colonies with increasing distance ($\rho$). In this figure it is assumed that $m_1 = 0.1$ and $\bar{m}_\infty = 4 \times 10^{-5}$. (From Kimura and Weiss 1964.)

district in Italy and by the help of computers, Cavalli-Sforza pioneered what may be called "demographic genetics."

Stimulated by the wealth of consanguinity information in this district, he and his collaborators developed a mathematical theory to predict the frequencies of various types of consanguineous marriages (Cavalli-Sforza, Kimura and Barrai 1966). The parameters that are necessary for the prediction are those concerning migration patterns, age distribution and similarity of mates in the *general population.* Also the population density is required. The agreement between the observed and expected frequencies was only fair, but with improvement of demographic information and refinement of the mathematical treatment, more satisfactory agreement can be expected (see also Cavalli-Sforza 1969). As an indicator for the breeding structure of populations, the probability of consanguineous marriages should have an important bearing on the genetics of human populations.

One of the essential parameters required for such a study is the distance of matrimonial migration, namely the distance between birth places of husband and wife. At present, not many reliable data are available on the frequency distribution of this distance. The data obtained from Parma Valley area (Cavalli-Sforza 1958) seem to be fitted well by a gamma distribution

$$f(r) = \frac{k^n}{\Gamma(n)} e^{-kr} r^{n-1}, \tag{27}$$

in which $r$ is the distance of matrimonial migration. Taking 0.625 km as the unit distance, the values of the constants turned out to be $k = 0.0324$ and $n = 0.0362$ for the plain area, and $k = 0.0155$ and $n = 0.1542$ for the mountainous area (Cavalli-Sforza 1962). The frequency distributions thus obtained have the property that the frequency at the zero distance is very high, since $f(0) = \infty$ for $n < 1$.

A similar study was recently undertaken by Yasuda (1968) for the Mishima district in Japan, a plain area including Mishima city and its four neighboring towns. For this purpose, he used "Koseki" at the city office, Koseki being a household record of a family nucleus with husband, wife and their children. In his study, particular attention was paid to clarifying the pattern of distribution at the neighborhood of zero distance. Figure 8.6 shows the observed frequency distribution for the matrimonial distance. It was found that the observed distribution in this area can be fitted by

$$f(r) = Cre^{-k\sqrt{r}}, \tag{28}$$

where $C = k^4/12$. With 1 km as the unit distance, $k = 2.85$ was obtained. Note that this distribution has quite a different form from the previous distribution obtained for the Parma district in that the frequency at zero distance is very low. In fact $f(0) = 0$ for this distribution. Clearly, more research on this subject is needed if we want to understand more fully the breeding structure of human populations.

Finally, we would like to discuss briefly some problems of mating systems with special reference to the rate of

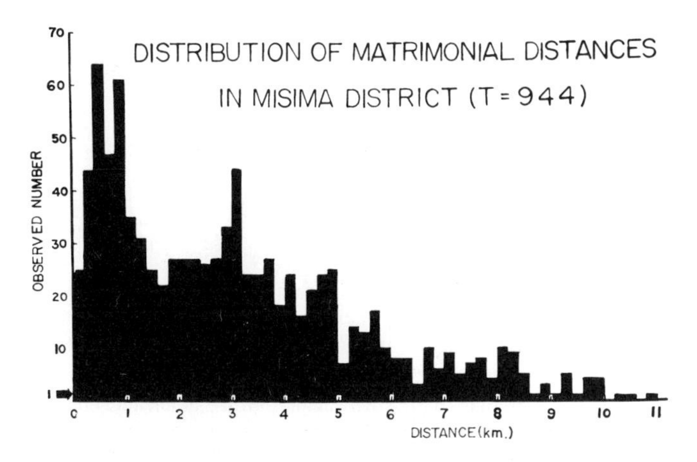


FIGURE 8.6. Frequency distribution for the matrimonial distance in the Mishima district. (From Yasuda 1968.)

decrease of heterozygosity in a finite population. In his classical paper on evolution of Mendelian populations, Wright (1931) showed that in a population of $N$ breeding individuals consisting of equal numbers of males and females, the rate of decrease of heterozygosity per generation is approximately $1/(2N)$ under random mating. With the same number of individuals, but under "maximum avoidance" of consanguineous mating, the corresponding rate becomes asymptotically $1/(4N)$ (cf. Wright 1951). Namely, the ultimate rate of decrease of heterozygosity is halved. The reason for the halving is not due primarily to the mating of less related individuals but is caused almost entirely by the fact that in the maximum avoidance system each parent produces exactly the same number of progeny. It may be natural, however, to infer from this that $1/(4N)$ is the minimum rate of decrease of heterozygosity that can possibly be attained with $N$ breeding individuals per generation, since common sense dictates that the more intense the inbreeding, the more rapidly heterozygosity is lost. Therefore, it was quite unexpected

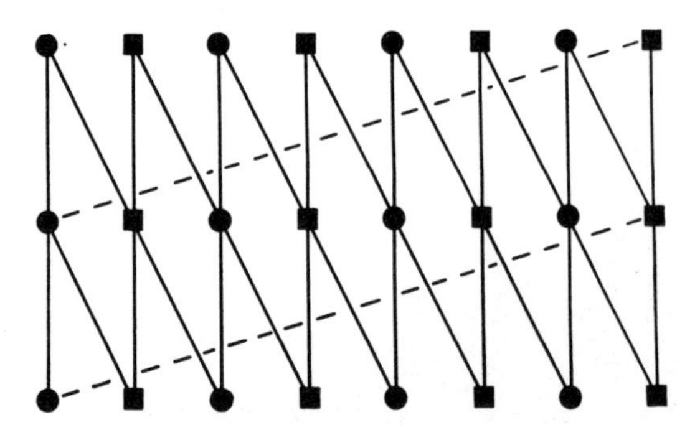

FIGURE 8.7. Circular mating with $N = 8$. More generally, circular mating consists of equal numbers of males and females arranged alternately so that each individual is mated with its neighbor. The last individual is mated with the first so that the system is visualized as circular. (From Kimura and Crow 1963b.)

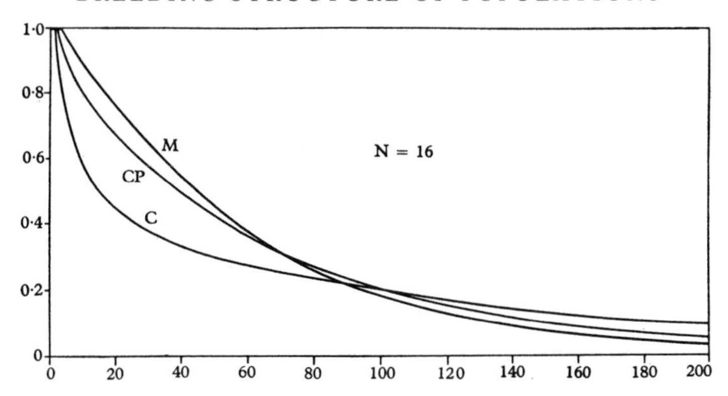


FIGURE 8.8. Decrease of heterozygosity with generations in a population of $N = 16$ for circular mating (C), circular pair mating (CP) and maximum avoidance of inbreeding (M). Ordinate: heterozygosity as a fraction of initial heterozygosity; abscissa: time in generations. (From Kimura and Crow 1963b.)

to find (Kimura and Crow 1963b) that under circular mating (Figure 8.7) the ultimate rate of decrease of heterozygosity is approximately $\pi^2/(2N + 4)^2$, which can be very much smaller than the corresponding rate under the "maximum avoidance" system. Note that circular mating consists of a circle of half-sib matings.

Compared with the maximum avoidance system, the circular mating system loses heterozygosity more rapidly in earlier generations, though eventually it loses heterozygosity at much lower rate (Figure 8.8).

The nature of such a seemingly paradoxical result was further pursued by Robertson (1964) and Wright (1965). Their studies have thrown much light on the problem of mating systems in relation to the amount of random drift and loss of heterozygosity. It is intriguing that more than 40 years had elapsed since the classical work of Wright on systems of mating (Wright 1921) before the nature of circular mating was discovered and clarified.

Ideas of the stepping stone model and the circular mating

system were recently combined by Maruyama (1970a) who investigated the "circular stepping stone model." In this model, the entire population consists of a circle of $2n$ colonies each having $N$ breeding individuals and exchanging individuals with adjacent colonies at the rate of $m$ per generation. Using this model, he has obtained the interesting result that the rate of decrease of heterozygosity is approximately $m\pi^2/(2n)^2$ if the rate of migration is so low that $m < n/(\pi^2 N)$. In this case, the rate of decrease of heterozygosity is proportional to the migration rate but independent of the colony size, so it can become very small if migration is very rare. On the other hand, if $m > n/(\pi^2 N)$, the rate of decrease of heterozygosity is approximately $1/(4nN)$ which is equal to the case in which the entire population of $2nN$ individuals forms a single panmictic population.

Maruyama also studied the number of neutral isoalleles maintained in a finite subdivided population of stepping stone structure, assuming that mutation to new and not preexisting alleles occurs at a rate $u$ per generation (Maruyama 1970c). This is an extension of the work of Kimura and Crow (1964) which we shall discuss in the next chapter.

A general result that he obtained from his study is

$$\bar{f} = \frac{(1-u)^2(1-f_0)}{2N_e[1-(1-u)^2]} \tag{29}$$

or approximately

$$\bar{f} = (1-f_0)/(4N_e u), \tag{30}$$

where $\bar{f}$ is the probability that two homologous genes randomly extracted from the whole population are identical by descent, while $f_0$ is the corresponding probability when they are extracted from the same subpopulation, and $N_e$ is the effective number of the whole population. It is assumed that random mating is practiced within each subpopulation.

Later it was found (Crow and Maruyama 1971) that formula (29) applies quite generally for any subdivided structure and any patterns of migration as long as the total population size and the breeding structure remain the same. This may be proved as follows.

Let $P$ be the probability of two randomly chosen homologous genes coming from the same subpopulation or colony, and let $Q$ be the conditional probability of their coming from the same individual given that they come from the same colony. If we denote by $f_1$ the probability that two homologous genes derived from two different colonies are identical by descent, we have, by definition

$$\bar{f} = Pf_0 + (1 - P)f_1. \tag{31}$$

If we consider the population in the next generation, we have

$$\bar{f}' = (1 - u)^2[P\{Q\left(\frac{1 + f_0}{2}\right) + (1 - Q)f_0\} + (1 - P)f_1]. \tag{32}$$

Here we assume that the population is monoecious. At equilibrium, $\bar{f}' = \bar{f}$, and if we substitute $\bar{f} - Pf_0$ for $(1 - P)f_1$ from (31) into (32), we obtain

$$\bar{f} = \frac{(1 - u)^2(1 - f_0)}{1 - (1 - u)^2}\left(\frac{PQ}{2}\right). \tag{33}$$

Since $PQ$ is the probability that two uniting gametes come from the same parent, as shown in Chapter 3, its reciprocal $1/(PQ)$ is equal to the inbreeding effective number of the population $N_{e(f)}$. Therefore (33) agrees with Maruyama's formula (29). Since (32) is concerned with the average relationship in the whole population, it is evident that his formula is valid for any pattern of migration as long as the whole population is closed and the breeding structure remains constant from generation to generation. It is probable that this formula will turn out to be useful for analyzing the observed genetic variability in structured populations.

CHAPTER NINE

# Maintenance of Genetic Variability in Mendelian Populations

Experimental and observational studies have made it increasingly clear that natural populations of sexually reproducing organisms such as man and *Drosophila* contain a large amount of genetic variability, ranging from conspicuous polymorphisms down to invisible molecular variations.

What then is the mechanism by which such variability is maintained? Few will disagree that this is one of the most important problems in population genetics.

It is clear that the ultimate source of genetic variation is mutation. However, mutation per se may not be sufficient to explain the wealth of variability maintained within a large outbreeding population.

Although no definite answer to this problem has yet been obtained, we shall endeavor in what follows to organize the observed facts and present theoretical models that may be pertinent to the solution of this problem.

First, we shall summarize the pertinent observations. There are a number of morphological variations that are genetically determined and some characters such as shell banding in land snails and mimicry patterns in butterflies have been analyzed. An important conclusion obtained is that a group of genes concerned with one trait are often organized into a gene complex or "super-gene" (cf. Sheppard 1969, Ford 1965).

Studies have also been made on "quantitative" characters such as height in man, bristle numbers in *Drosophila* and so on. The high correlation between parent and offspring and the effectiveness of artificial selection suggest that these variations are largely determined genetically and that much of the variance is additive. Recent analyses of the sternopleural bristles in *Drosophila* and the heading time in wheat suggest that variations are mostly determined by segregation at only 10 or less of loci and that the genes involved act additively (cf. Robertson 1967). It is possible that the number of loci having very small effects is much larger.

Chromosome polymorphisms are found to be abundant in *Drosophila,* where the large salivary gland chromosomes can be studied easily, and in *Trillium* (cf. Haga and Kurabayashi 1953), where differential reactivity of chromosome segments makes identification precise. In most organisms these techniques are not available, and so the possibility of detection of inversions or translocations is much more limited.

Great progress in the analysis of fitness traits, especially of viability, has been made in *Drosophila* through the use of special chromosomes that suppress crossing over and are marked with dominant mutants. It enables us to measure the relative effect on viability of chromosomes homozygous or heterozygous in comparison with a particular marker chromosome. The following table shows homozygous loads in the second and the third chromosomes of *Drosophila melanogaster* obtained by *Curly Moire* technique (cf. Temin et al. 1969). From the table it may be seen that, on the average, one out of four second or third chromosomes has a lethal gene and an average chromosome has some detrimental genes that are equivalent to $0.15 \sim 0.18$ lethals. They may consist of many detrimentals with mild effects or a smaller number of detrimentals with severe effects.

TABLE 9.1. Homozygous loads for the second and the third chromosomes of *Drosophila melanogaster*. (From Temin et al. 1969.)

| | Second chromosome | Third chromosome |
|---|---|---|
| Mild detrimentals | 0.095 | 0.122 |
| Severe detrimentals | 0.062 | 0.053 |
| Total detrimentals | 0.157 | 0.175 |
| Lethal | 0.247 | 0.240 |

The above results refer to the homozygous load of the equilibrium population. On the other hand, Mukai (1964) studied the effect of newly arising mutations on viability. According to his results, the accumulation of lethal mutants and detrimental loads in the second chromosome of *D. melanogaster* are respectively about 0.005 and 0.0039 per generation.

The mechanism of maintenance of lethal genes in the population is now reasonably well understood, and we can conclude that the great majority of them are maintained by the balance between mutation and selection. We know also that the selection is mainly in the heterozygous state, the lethal heterozygotes having on the average a few percent lower fitness than the normal homozygotes (Crow and Temin 1964, Mukai 1969a, Nei 1968a). There are still those who claim that a majority of "recessive" lethals are heterotic in their own genetic background with some 10% overdominance. If this claim were correct, equilibrium frequency of each lethal would be nearly 10% and an average second chromosome would contain several dozen lethals, contrary to observational facts. The same conclusion also applies to the detrimentals reducing viability.

An interesting fact is that the estimated number of loci capable of producing lethal mutants is only several hundred for the second chromosome of *D. melanogaster* (Muller 1950 and Wallace 1950). At the rest of the loci, a "point mutation" may not be able to lead lethality. The number of loci which cause detrimental effects on viability must be much larger and may reach at least $10^4$ for the second chromosome of *D. melanogaster* if we use Mukai's estimate of mutation rate for viability polygenes (some 14% per generation) and if we assume the mutation rate of $10^{-5}$ per locus. This figure may represent the total number of conventional genetic loci. On the other hand, Muller's method provides only a minimum estimate of the number of lethal producing loci and it may be a serious underestimate if the mutation rate varies greatly from locus to locus. Lifschytz and Falk (1969) located 34 cistrons where lethals were produced in a region of the *X* chromosome $1.5 \sim 2.5$ map units in length. Since the total map length is $250 \sim 300$, this corresponds to something like 5,000 lethal producing loci per haploid set of chromosomes. So the higher rate of mutation for detrimentals may be a higher rate of mutation per locus, rather than a larger number of loci, or both.

Although there is considerable room for uncertainty about the number of loci involved, one fact is very clear; namely, that the total mutation rate is much higher for mildly deleterious mutants than for lethals.

We should like to emphasize here that the relative viability, especially as measured using marker chromosomes, is not the only important factor for the determination of fitness. Fertility or fecundity may be as important as viability, although its measurement is much more difficult. It may be that, just as with viability genes, the total mutation rate for mutants causing a slight reduction in fertility or fecundity is 10 or more times as high as that for complete sterility. An interesting additional finding is that most mutants causing sterility do so in only one sex. A majority of such mutant genes may be kept in the population by mutation-selection balance.

Now, we switch our attention to the genetic variation at the enzyme level that can be observed by the techniques of electrophoresis and immunology. Recent developments in the former technique have made it possible to detect iso-allelic polymorphisms in various organisms, and it has turned out that the genetic variability is much more extensive than was previously thought, although this possibility was foreseen and discussed by Kimura and Crow (1964). According to Lewontin and Hubby (1966) the average heterozygosity is 12% and the proportion of polymorphism is 30% for 18 randomly chosen loci in *D. pseudoobscura.* Later reports by various investigators have confirmed these results. Similar results have been reported on the enzymes of human blood by Harris (1966) and Lewontin (1967). The heterozygosity is 16% and the proportion of polymorphism is 36% for 33 loci sampled. Thus, large out-breeding populations like *Drosophila* and man contain tremendous genetic variability. Moreover, we must note that, by the method of electrophoresis, only $1/4 \sim 1/2$ of the amino acid changes can be detected.

If we naively estimate the total number of genes in *Drosophila* by dividing the total number of nucleotide pairs in the genome (ca. $2 \times 10^8$) by the average number of nucleotide pairs per cistron (ca. 500) we get $4 \times 10^5$. So, the actual number of heterozygous as well as polymorphic loci by this method is enormous. The gene number estimated in this way is a magnitude higher than the earlier estimate of the number of conventional genes by Muller and others (cf. Muller 1966) based on mutation rates, giving $10^4$ at most. This discrepancy, we believe, reflects the fact that a large fraction of DNA in the genome of higher organisms is not informational (Ohta and Kimura 1971c). In such non-informational part of DNA, polymorphism is expected to occur in higher frequency than in cistrons, since practically all mutations are selectively neutral, that is, there is no selective constraints in such part of DNA.

We shall now discuss some possible mechanisms responsible for such a great amount of genetic variability. The main alternative hypotheses that have been actively discussed recently are selection (mainly overdominance) versus mutational production of neutral or nearly neutral isoalleles. There are many other possible causes which may lead to intermediate gene frequencies at equilibrium, but they are not usually thought to be as important as the above two.

We first consider the model of neutral and nearly neutral isoalleles first suggested by Kimura and Crow (1964) followed by more detailed study by Wright (1966) and Kimura (1968b). It is now well known that mutation is caused by changes in DNA base pairs, such as substitution, deletion and rearrangements. Therefore, each gene can take an astronomical number of allelic states and some of these may have a very similar effect on fitness. Assuming that each new mutant is an allelic state not preexisting in the population, the upper limit to the number of different isoalleles maintained within a finite population can be estimated.

Consider a particular locus and let $u$ be the average rate of neutral mutation, such that in the population of actual size $N$, $2Nu$ new mutants will be produced in each generation. Let $N_e$ be the effective size of the population. Then it can be shown that, at equilibrium, the probability that an individual is homozygous is

$$F = \frac{1}{4N_e u + 1}. \tag{1}$$

Thus, the heterozygosity, $H$, is $1 - F$, or

$$H = \frac{4N_e u}{4N_e u + 1}. \tag{2}$$

The important quantity is $4N_e u$, and if this is larger than unity the heterozygosity is higher than ½. Kimura and Crow (1964) defined the reciprocal of the sum of the squares of allelic frequencies as the effective number of alleles in the population. In the present case $n_e = 1/F$ and we have

$$n_e = 4N_e u + 1. \tag{3}$$

This is almost always smaller than the actual number of alleles contained in the population, for rare alleles contribute little to heterozygosity.

If the possible number of allelic states is $K$ rather than infinity, we have, assuming that every allele mutates to the remaining $(K-1)$ alleles with the rate $u/(K-1)$ per generation,

$$n_e = \frac{4N_e u\left(\frac{K}{K-1}\right) + 1}{4N_e u\left(\frac{1}{K-1}\right) + 1} \tag{4}$$

as a good approximation (Kimura 1968b, Crow and Kimura 1970a, p. 453). Generally, formula (1) is valid whenever $2N_e u^2 \ll 4N_e u + 1 \ll K$. Note that for each nucleotide site, $K = 4$, so the heterozygosity is

$$H = \frac{4N_e u}{1 + 16N_e u/3}, \tag{5}$$

where an appropriate value for $u$ is probably $10^{-8}$ or less.

Since the effective number of alleles ($n_e$) is usually much smaller than the total number of alleles actually contained in a population, it may be convenient to introduce another quantity, $n_a$, which has been termed the average number of alleles. This is the average of the number of alleles actually contained in the population. In the model of $K$ allelic states with equal mutation rates in all directions, it can be shown that

$$n_a = C \int_{1/(2N)}^{1} (1-x)^{\alpha-1} x^{\beta-1}\, dx \tag{6}$$

where

$$C = \frac{K\Gamma(\alpha+\beta)}{\Gamma(\alpha)\Gamma(\beta)} \tag{7}$$

in which $\alpha = 4N_e u$ and $\beta = \alpha/(K-1)$. It is assumed that all the alleles have the same fitness and that the rates of mutation from any allele to any other are the same. The total rate of mutation from one allele to all others is $u$. At the limit of $K \to \infty$, we have $\beta = 0$ and $C = \alpha = 4N_e u$. Figure 9.1 shows the relationship between the number of alleles ($n_e$ or $n_a$) and $N_e u$.

An important question, other than that of heterozygosity, is what is the probability that a particular locus is polymorphic? As shown in the Appendix (see A2.10), if the number of allelic states is very large, the probability that a population is monomorphic with respect to a particular neutral locus is

$$P_{\text{mono}} = q^{4N_e u}, \tag{8}$$

where $q$ is a small positive value such that we call a population "monomorphic" if the total frequency of "variant" alleles is $q$ or less. In other words, we regard the population essentially monomorphic if the frequency of the allele that happens to predominate in the population is higher than $1-q$. The value of $q$ is arbitrary, but a reasonable value may be $q = 0.01$. From formulae (2) and (8), we have the following formula for the probability of polymorphism:

$$P_{\text{poly}} = 1 - q^{H/(1-H)}. \tag{9}$$

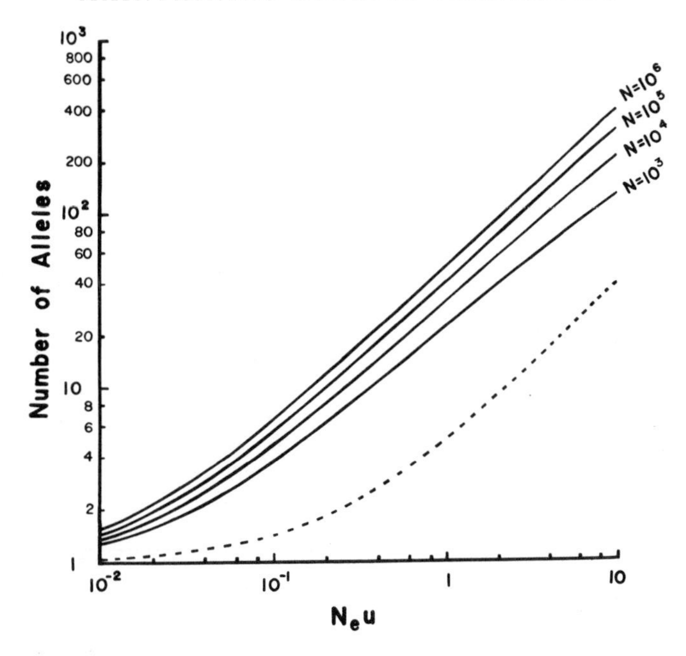


FIGURE 9.1. Relationship between the number of alleles and $N_e u$ (product of the effective population number and the mutation rate). The solid lines give the average number of alleles ($n_a$) at four levels of population number ($N$), and the broken line gives the effective number of alleles ($n_e$), which is independent of $N$. (From Kimura 1968b.)

Figure 9.2 illustrates the relationship between $\bar{H}$ and $P_{\text{poly}}$ for $q = 0.01$ and 0.05 together with the observed values (from Table 3 of Selander et al. 1970). The agreement between theoretical and observed values is satisfactory.

Another model that may be pertinent to the investigation of heterozygosity at the molecular level is the one employed by Kimura (1969b). In this model, we consider the individual nucleotide site rather than the conventional genetic locus (cistron) as the unit of mutation, and assume that the total number of nucleotides making up the haploid genome is so large while the mutation rate per site is so

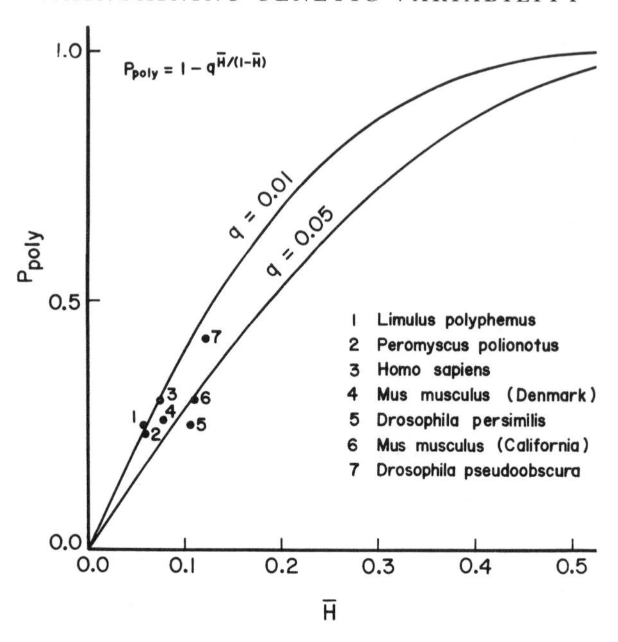


FIGURE 9.2. Relationship between the probability of polymorphism ($P_{\text{poly}}$) and the average heterozygosity ($\bar{H}$) for two levels of $q$(0.01, 0.05), together with observed values represented by dots.

low that whenever a mutant appears, it represents a mutation at a previously homallelic site.

Let $\nu_m$ be the average number of sites in which new mutants appear in each generation in the entire population. Then it can be shown that if the mutants are neutral, the total number of heterozygous nucleotide sites per individual at equilibrium is

$$H_n \approx 2\nu_m \frac{N_e}{N}, \tag{10}$$

and the total number of segregating sites within the population is

$$I \approx 2\nu_m \left(\frac{N_e}{N}\right) \{\log_e (2N) + 1\}. \tag{11}$$

Now, let us examine the adequacy of the neutral mutation theory to explain the isoallelic variations observed by electrophoretic methods. Our main idea is to regard protein polymorphisms as a phase of molecular evolution (Kimura and Ohta 1971) as discussed in Chapter 2. We have seen that the average rate of amino acid substitutions is about 1.6 paulings ($1.6 \times 10^{-9}$ per year per amino acid). If most of the amino acid substitutions are selectively neutral, this should be equal to the mutation rate for neutral isoalleles. Therefore, if the average cistron corresponds to 300 amino acids, the neutral mutation rate per locus is $u = 1.6 \times 10^{-9} \times 300 \approx 5 \times 10^{-7}$ per year. This is much smaller than the conventional mutation rate per locus per generation of $10^{-5}$ since the average generation time of most mammals is at most a few years. This suggests that the neutral mutations constitute a rather small fraction of the total mutations. It is likely that this fraction differs from locus to locus depending upon the functional requirement of each protein.

Shaw (1965) has estimated that only $1/2 \sim 1/4$ of the mutants involving amino acid changes lead to detectable electrophoretic variations. This figure should also apply to neutral mutants, but it gives an overestimate if the fraction of neutral mutants is less among mutants that can be detected by electrophoresis than among those that cannot be so detected. Also, it is possible that the changes that can be detected by electrophoresis are restricted to amino acids that are exposed to the surface of the molecule and, therefore, the above figure is an overestimate. For these reasons, we take $u = 10^{-7}$. For species such as the mouse with possibly two generations per year, the mutation rate per generation for such mutants is probably half as large, while for man it should be some 20 times as large. Nevertheless, the average heterozygosity observed for enzyme polymorphisms is about the same for mouse and man, being respectively 9.4% (Selander et al. 1969) and 7.4% (Harris 1969), both giving roughly $4N_e u = 0.1$. Then the effective population number that satisfies this relation is $N_e \approx 0.5 \times 10^6$ for the mouse and $N_e \approx 1.3 \times 10^4$ for man. The effective number here refers to the species or subspecies in the course of evolution.

The above discussion leads us to conclude that mutation seems to be sufficient for the production of observed heterozygosity in large panmictic species such as *Drosophila, Mus* and man. Protein polymorphisms can thus be regarded simply as the transient phase of molecular evolution. In this view, the protein polymorphism and the molecular evolution are not two separate phenomena, but merely two aspects of a single phenomenon caused by random frequency drift of neutral mutants in finite populations. Then, we should expect that not only genes in "living fossils" have undergone as many DNA base (and therefore amino acid) substitutions as corresponding genes in more rapidly evolving species as predicted by Kimura (1969b), but also that they are equally polymorphic and heterozygous at the protein level. Recent study by Selander et al. (1970) on the variation of the horseshoe crab at the protein level appears to support this view.

The apparent objection to the theory, however, seems to lie not in the level of heterozygosity per se but in the nature of variation. As pointed out by Robertson (1968), either one of the two alternatives must happen: Either many different alleles are segregating in large populations or a small number of different alleles are segregating in different, isolated small populations, assuming that a large fraction of the mutation at a cistron is selectively neutral. The chromosomal variation in *Trillium* may be a good example of this type of variation (Haga 1969). Many chromosome types can be distinguished with respect to heterochromatic segmentation patterns in this species. Among small isolated populations, different chromosome types are fixed in the homozygous state, while in large populations, several chromosome types are segregating in Hardy-Weinberg zygotic proportions. This example appears to be in sharp contrast to the inversion polymorphisms in *Drosophila* in which there are clear cases of strong overdominance.

Recent observational studies (Prakash et al. 1969, O'Brien and MacIntyre 1969, Selander et al. 1969) using *Drosophila* and *Mus* have shown a rather constant distribution of the same set of alleles in different local populations far apart from each other. These authors believe that their observational results suggest the existence of some kind of balancing selection. But, is selection absolutely necessary for the constant distribution of isoalleles among different subpopulations?

In our view, the constant distribution of the same set of alleles only indicates sufficient migration between local populations to make the entire species or subspecies practically panmictic. Recently, Maruyama (1970a, b, c, 1971) has carried out extensive mathematical analysis of the stepping stone model of finite size. He has worked out the exact relationship between local differentiation of gene frequencies and the amount of migration. His results show that in the two dimensional stepping stone model, if $N$ is the effective size of each colony and $m$ is the rate at which each colony exchanges individuals with four surrounding colonies per generation, then marked local differentiation is possible only when $Nm$ is smaller than unity (assuming a large number of colonies arranged on a torus). This is a very severe restriction for migration between colonies because the number of individuals which each colony exchanges with surrounding colonies must be less than one on the average per generation, irrespective of the size of each colony. For the model of continuous distribution of individuals over an area, this condition is equivalent to

$N_\sigma < \pi$, where $N_\sigma$ is the average number of individuals within a circle of radius $\sigma$, the standard deviation of the distance of individual migration in one direction per generation. If, on the other hand, there is more migration, the whole population tends to become effectively panmictic. The transition from marked local differentiation to practical panmixis is very rapid for the distribution over an area, and it can be shown that if $N_\sigma > 12$, the whole population behaves as if it were a single panmictic population.

This means that when two or more alleles happen to be segregating within a species, their frequencies among different localities far apart from each other are nearly the same (Kimura and Maruyama 1971). For animals with separate sexes, it is expected that at least several individuals of males and females usually exist within a circle of radius $\sigma$ and $N_\sigma > 12$ is therefore met almost always by widely distributed and actively moving animals. Maruyama also showed that when isolation is more complete and different alleles tend to fix in different local populations, they are connected by zones of intermediate frequencies. The overall pattern then mimics a gene frequency cline due to selection, even if alleles are in fact neutral.

The agreement of the neutral mutation theory with the observed facts on polymorphisms and the molecular evolution does not by itself constitute a final proof. Such an agreement is, so to speak, a necessary condition but not a sufficient condition for the hypothesis to be finally established. Therefore we shall discuss some alternative hypotheses.

Let us examine the extreme model which assumes multiplicative overdominance in most, if not all, of the isozyme polymorphisms. This model is inadequate, for the population must suffer an intolerable genetic load due to segregation, as we have already seen in Chapter 4. Thirty percent polymorphism means 3,000 polymorphic loci, if the total number of loci is $10^4$. In order to avoid such a heavy load, King (1967), Sved et al. (1967) and Milkman (1967) introduced truncation type selection, but this has the shortcoming that it does not give the observed pattern of inbreeding depression (Chapter 4). Furthermore, it is highly doubtful that natural selection mimics artificial selection in such a way that the total number of heterozygous loci for each individual is counted and the individual is culled if the number of heterozygous loci is less than a certain critical number.

Another important point which is often forgotten in many discussions of balanced polymorphism is that overdominance is not effective in self-fertilizing populations. In many plants, partial self-fertilization is very common. Isozyme polymorphisms in plants have not been so well investigated as in animals and we cannot generalize from only a few examples. However, esterase, phosphatase and anodal peroxidase show considerable polymorphism in *Avena barbata* (Marshall and Allard 1971). If we let $S (0 \leqslant S \leqslant 1)$ be the proportion of self-fertilization, it can be shown (Appendix A4) that for a non-trivial equilibrium to be stable, the proportion of self-fertilization must be such that

$$S < \frac{2s_2(1 - s_2)}{s_1 + s_2 - 2s_1s_2}, \tag{12}$$

where $s_1$ and $s_2$ are the selection coefficients against the two homozygotes (assuming $s_1 \geqslant s_2 > 0$). The formula indicates that unless the heterozygote is more than twice as fit as either homozygote, the amount of self-fertilization must be restricted. For example, a pair of alleles with $s_1 = 0.5$ and $s_2 = 0.1$ can be maintained in balanced polymorphism only in a population in which the proportion of self-fertilization is less than 36%. The restriction given in (12) suggests that overdominance cannot be a major mechanism for polymorphism in predominantly self-fertilizing organisms, unless the overdominance is very strong.

Thus strict overdominance is not theoretically adequate to explain most of the isoallelic variabilities. Therefore we shall next examine the model in which only a small fraction of loci are selected (Ohta and Kimura 1971b). It may be that some part of genetic variabilities are selected and that overdominance or some other form of balanced selection exists among them. The mutant for sickle cell hemoglobin is a well known example of overdominance. The $\beta^s$ allele has a selective disadvantage by causing severe anemia in the homozygote, but gains advantage by producing resistance to malaria in $\beta^s\beta$ heterozygotes. We believe that this type of ambivalent gene action (Huxley 1955) is the main cause of overdominance. In other words, each of the two alleles has an advantage through a different aspect of fitness, resulting in overdominance in overall fitness. It is known that several kinds of multimer enzymes are formed in heterozygotes and that they sometimes show complementary function. It is possible that differential activity of multimers will sometimes result in the heterozygote advantage (Fincham 1966).

In the following discussion, we assume that most of isoallelic variabilities are selectively neutral, but that there also are a number of truly overdominant loci sparsely distributed (as compared with neutral loci) over the entire genome. For this model, the effect of linkage disequilibrium as discussed in Chapter 7 becomes most important. In particular, if a selectively neutral mutant allele is more or less tightly linked to overdominant loci, "associative overdominance" will be developed at the neutral locus and the heterozygosity might sometimes be enhanced.

To be more specific, suppose 10% of isoallelic loci are kept polymorphic by overdominance and the rest are selectively neutral. If we take 30% as the proportion of polymorphic loci and $10^4$ as the total number of loci in *Drosophila,* there are 300 overdominant loci. Three hundred overdominant loci each with genuine but slight overdominant effect such as one percent heterozygote advantage including fertility and fecundity may be acceptable from the point of view of genetic load and inbreeding depression. This would give about one overdominant segregating locus on each chromosome at the interval of one centimorgan (one crossover unit) while neutral polymorphisms exist 10 times as much.

We have already shown in Chapter 7 (see Figure 7.1) that if there are $n_1$ overdominant loci on the left and $n_2$ overdominant loci on the right of an intrinsically neutral locus with $c_0$ recombination fraction between adjacent loci, and if $s$ is the selection coefficient against either homozygote at each overdominant locus, then the apparent heterozygote advantage at the neutral locus is given approximately by

$$s' = \frac{s}{4N_e c_0}(2\gamma + \log n_1 + \log n_2), \qquad (13)$$

where $\gamma \approx 0.58$. The effect of such associative overdominance depends on the quantity $N_e s'$, but as seen from formula (13), this quantity is independent of $N_e$. For example, if $n_1 = n_2 = 50$ and $c_0 = 0.01$, we have $N_e s' = 2.25$ if $s = 0.01$ and $N_e s' = 4.5$ if $s = 0.02$. At present we are not sure of the correspondence between associative overdominance and the retardation factor of Robertson (1962). But associative overdominance will retard the fixation of alleles with intermediate frequencies.

There is a very interesting property of associative overdominance, as stated above, that $N_e s'$, the product of effective population size and the apparent heterozygote advantage, is independent of the population size. Also, if we change the number of overdominant loci such that the total effect on the given segment does not change (for example, $2n$ loci each with homozygous disadvantage $s/2$ instead of $n$ loci each with $s$), the amount of associative overdominance changes relatively little. Thus, if the inbreeding depression of fitness in competitive ability is about 50% at the level of inbreeding coefficient $f = 0.25$ and 90% at $f = 0.886$ as observed in *Drosophila melanogaster* by Latter and Robertson (1962), the previous example of 300 overdominant loci with $1 \sim 2\%$ heterozygote advantage is a realistic one.

So far we have concentrated our discussion on neutral mutation and overdominance. There are, however, numerous other possible factors which may be responsible for the maintenance of genetic variability, such as frequency-dependent selection, heterogeneous environment and meiotic drive. These may certainly be important for some loci, but we do not know how universal they are, especially in relation to numerous isoallelic variabilities.

Finally, let us consider the relationship between isoallelic variabilities and the viability polygenes. As pointed out by Mukai (1968) only some 10 viability polygenes are present in the average second chromosome of *D. melanogaster*. This is much too small a number as compared with the isozyme polymorphisms. Also, the correlation in fitness between homozygotes and heterozygotes shows that they are nearly semidominant (Mukai 1968). Therefore, we believe that there is very little relationship between Mukai's viability polygenes and isozyme variabilities.

We conclude that the extended form of the classical hypothesis can explain the maintenance of the majority of genetic variabilities. That is, the majority of lethals and detrimentals are maintained by the balance between mutation and selection, while the majority of isoalleles are maintained by the balance between mutation and random drift. The balancing selection would probably operate on a small fraction of the total variabilities. In addition, a small fraction of polymorphisms represents a transient phase of mutant substitution by natural selection, and they are particularly important for the adaptive evolution of the species.

## CHAPTER TEN

# The Role of Sexual Reproduction in Evolution

The origin of sexual reproduction must be very old, as may be inferred from the fact that the lowest of living things such as bacteria and viruses can undergo gene recombination through sexual processes. Furthermore, sexual reproduction seems to be so widespread that there appear to be no major groups of organisms that are entirely devoid of this means of reproduction.

On the other hand, as a means of increasing the population number, this may not be the best method. Asexual reproduction by budding or cloning can do the job much more efficiently and safely.

Then what is the advantage of sexual reproduction, if this has been so widely adopted throughout the evolutionary history? It is usually claimed that the advantage of sexual reproduction consists in its ability to produce enormous number of genotypes through recombination out of a relatively small number of segregating loci. Such a view is still held by many authorities in evolution theory (see, for example, Tax and Callender 1959, pp. 114–115).

Although the total number of potential combinations is very great, this number may equally be attained by mutation without recombination, if enough time is allowed. Furthermore, the actual number of combinations realized in each generation is limited by the population size. Also, gene combinations are broken up by recombination as readily as they are produced.

The greatest advantage of sexual reproduction, in our opinion, is its ability to bring together in one individual advantageous mutant genes that occur in different individuals, thereby *enhancing the rate of evolution,* as first pointed out independently by Fisher (1930b) and Muller (1932). Under asexual reproduction, on the other hand, two advantageous mutants can be incorporated into the population only when the second advantageous mutant is produced in a descendant of the individual carrying the first advantageous one. Figure 10.1 illustrates these points especially for a large population, where the advantage of sexual over asexual reproduction is manifest.

Quantitative treatment of comparing the rate of evolu-

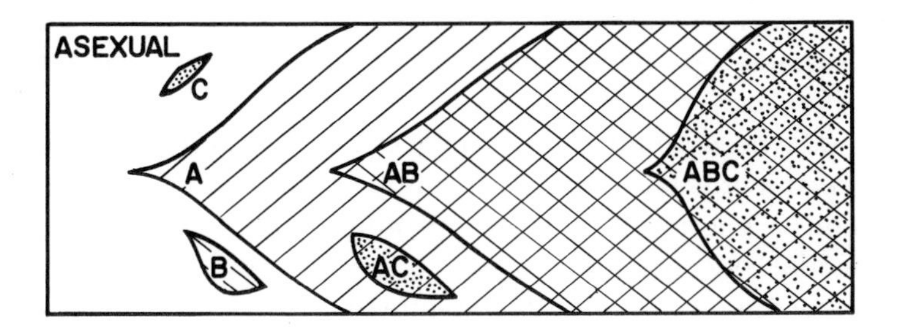


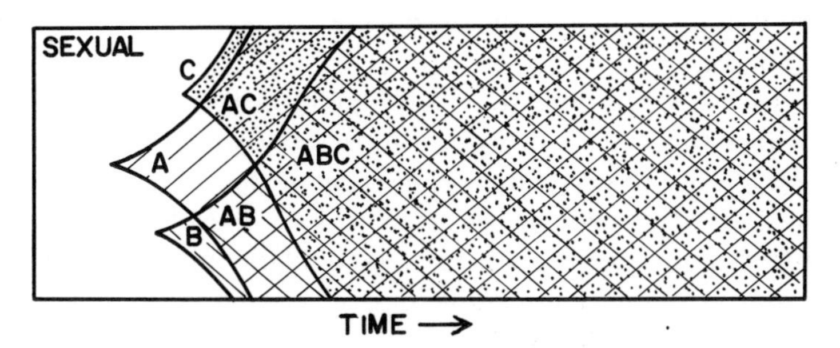


FIGURE 10.1. Diagrams illustrating different patterns of the processes of gene substitution in asexual and sexual populations both assumed to be large. (Slightly modified from Crow and Kimura 1965.)

tion under sexual and asexual population was made by Muller (1958) and later it was slightly improved by Crow and Kimura (1965). Their treatment, as well as Fisher-Muller theory, was criticized by Maynard Smith (1968b) who tried to show, based on a counter example involving two segregating loci, that sexual reproduction confers no advantage if the genetic variance is generated by mutation in a uniform environment. However, it was shown by Crow and Kimura (1969) that his criticism does not invalidate the essential part of the theory, though it does invalidate a very minor part of Muller's original argument which contended that asexual reproduction has additional disadvantage since different beneficial genotypes must compete with each other, each interfering with the evolution of the others, whereas under sexual reproduction the different advantageous mutants can combine and increase without hindrance.

Maynard Smith also claims that Crow and Kimura's treatment is based on the assumption that favorable mutations are unique events (an assumption that they have not made). According to him, in a large population and under recurrent mutation, double-mutant individuals already exist in the population before natural selection operates on them, so that their production by recombination is unnecessary.

In understanding the real significance of the Fisher-Muller theory, it is important to realize that advantageous mutations are very rare, yet because of their slow increase by natural selection, in the actual process of evolution, gene substitution goes on at many loci in a large population. In order to attain a comparable rate of gene substitution in an asexual population with that in a sexual one, it is necessary to start with multiple-mutant individuals, each carrying many advantageous mutant genes at the same time, which can occur only with a vanishingly small probability by mutation alone.

The treatment by Crow and Kimura (1965) is deterministic, but to be more accurate, a stochastic treatment is desirable.

In the following, we shall present a semi-stochastic version of their theory, taking account of the fact that a majority of individual mutants are lost by chance from the population even if they are advantageous in natural selection (see Chapter 1). The treatment, to be sure, is applicable irrespective of whether favorable mutations are recurrent or unique, as long as they are rare.

Let us consider first a population in which recombination is completely absent (i.e. an asexual population). Let $2v$ be the rate of occurrence per individual per generation of favorable mutations, and let $s$ be the average selective advantage of a favorable mutant. Throughout the treatment we consider only favorable mutants, disregarding deleterious and neutral ones. We shall denote by $g$ the average number of generations between the occurrence of a favorable mutation and the occurrence of another favorable mutation in the descendant of the first, considering only those "lucky" mutants that escaped accidental loss in early generations.

The effective rate of occurrence of such lucky mutations may be obtained approximately by multiplying $2v$ with the probability of fixation (see Chapter 1), which is $2s(N_e/N)$ if $s$ is small but $N_e s$ is large, where $N$ and $N_e$ are respectively the actual and effective population sizes. We assume that the offspring distribution is the same in sexual and asexual cases. Thus, we shall call

$$U = 2v \cdot 2s\left(\frac{N_e}{N}\right) = \frac{4svN_e}{N} \tag{1}$$

the "effective rate" for beneficial mutations per individual per generation.

From the deterministic theory, the frequency ($p_t$) of a beneficial mutant at time $t$, given that it is $p_0$ at the start, is

$$p_t = \frac{p_0}{p_0 + (1 - p_0)e^{-st}}. \quad (2)$$

The appropriate value for $g$ may then be obtained from

$$1/U = \int_0^g N p_t \, \mathrm{d}t, \quad (3)$$

since $1/U$ is the number of individuals such that one lucky favorable mutant is expected to occur, and since, by definition, $g$ is the number of generations during which another favorable mutant starts to increase.

From (3), putting $p_0 = 1/N$, we get

$$g = \frac{1}{s} \log_e \{N(e^{1/4N_e v} - 1) + 1\} \quad (4)$$

which corresponds to formula (5) of Crow and Kimura (1965).

On the other hand, under sexual reproduction, by virtue of free recombination, all the beneficial mutations that escaped accidental loss in early generations are incorporated. Namely, $Ng2v$ beneficial mutations occur during $g$ generations and approximately $Ng2v \cdot 2s(N_e/N) = 4N_e svg$ mutants are incorporated into the population.

Therefore, the ratio of the evolutionary rates in terms of gene incorporation between a sexual and an asexual population is $4N_e svg : 1$ or

$$R = 4N_e v \log_e \{N(e^{1/4N_e v} - 1) + 1\}. \quad (5)$$

Since

$$K = 4N_e vs$$

is the rate of gene substitution due to natural selection of advantageous mutants in a sexual population, the above formula (5) is also expressed as

$$R = \frac{K}{s} \log_e \{N(e^{s/K} - 1) + 1\}. \tag{6}$$

The following are some examples:

If $s = 10^{-2}$, $K = 3 \times 10^{-2}$ and $N = 10^6$, we have $R \approx 5.6$.

If $s = 10^{-3}$, $K = 10^{-2}$, $N = 10^7$, we have $R \approx 138$.

To be sure, the present treatment is a crude approximation and a more exact treatment may be required, particularly if equation (6) gives a large value.

Summing up, the advantageous effect of sexual over asexual reproduction in speeding up evolution is greatest if the population is large, the rate of gene substitution by natural selection is high, and the individual effect of advantageous mutants is small.

On the other hand, recombination under sexual reproduction is a definite disadvantage rather than an advantage if evolution is taking place mainly by bringing together individually deleterious but collectively beneficial mutants (epistasis).

Meanwhile, that asexual reproduction is often very efficient in multiplying individuals can be seen by the success of some perennial plants that reproduce exclusively by clonal multiplication, covering a very wide area. They are, so to speak, enjoying ephemeral luxury in evolution.

Finally, we would like to close this chapter by presenting a remarkable example to show Fisher's great insight on the problem. In his 1930 book, Fisher made the following prediction (p. 123):

"On this view, although asexual reproduction might be largely or even exclusively adopted by particular species of sexual groups, the only groups in which we should expect sexual reproduction never to have been developed, would be those, if such exist, of so simple a character that their genetic constitution consisted of a single gene."

After the many years since this was written, the development of bacterial and viral genetics has finally uncovered that even these lower organisms do undergo sexual recombination and that they indeed have more than one gene. Sexual reproduction has played a very important role in speeding up evolution in the past, helping to produce man before the sun in our solar system burns out.

Muller (1960) once said that the gene, through the long course of evolution, has finally found itself by producing man. We would like to add that man, by producing genetics, has finally started to understand the biological meaning of himself.

# Mathematical Appendix

## A1. *The Diffusion Equation Method*

The fundamental quantity which is used to represent the genetic composition of a Mendelian population is the gene frequency, or the proportion of a particular allelic gene in the population.

The most powerful method that has so far been developed to treat the process of change in gene frequency as a stochastic process (chance events proceeding with time) makes use of diffusion equations, in particular, the Kolmogorov forward and backward equations—the method now known as "diffusion models" (Kimura 1964). This is essentially an approximation which takes advantage of the fact that each gene is a self-reproducing entity and therefore the gene frequency in any reasonably large population changes almost continuously with time.

Let us suppose that a pair of alleles, $A_1$ and $A_2$, are segregating at a locus, and let $\phi(p, x; t)$ be the probability density that the frequency of $A_1$ becomes $x$ at time $t$ (conveniently measured with one generation as the unit length of time), given that it is $p$ at the start ($t = 0$). Then $\phi$ satisfies the following partial differential equation which is known as the Kolmogorov forward or the Fokker-Planck equation.

$$\frac{\partial \phi}{\partial t} = \frac{1}{2} \frac{\partial^2}{\partial x^2} (V_{\delta x} \phi) - \frac{\partial}{\partial x} (M_{\delta x} \phi), \qquad \text{(A1.1)}$$

where $M_{\delta x}$ and $V_{\delta x}$ are the mean and the variance of the change in $x$ per generation. More precisely, if $\delta x$ is the amount of change in gene frequency $x$ during a short time interval of length $\delta t$, then $M_{\delta x}$ and $V_{\delta x}$ are respectively the limits of $E_\delta(\delta x)/\delta t$ and $E_\delta\{(\delta x)^2\}/\delta t$ at $\delta t \to 0$, where $E_\delta$ is the operator of taking expectation with respect to $\delta x$. Equation (A1.1) is fundamental in treating the process of change in gene frequency in finite populations, and it was first introduced in population genetics in this form by Wright (1945a).

Several methods of derivation of equation (A1.1) are now available that differ in mathematical rigor and sophistication. We give here the one which we think most straightforward. Let $g(x, \delta x; t, \delta t)$ be the probability density that the frequency of $A_1$ changes from $x$ to $x + \delta x$ during the short time interval between $t$ and $t + \delta t$. Then

$$\phi(p, x; t + \delta t) = \int \phi(p, x - \delta x; t) g(x - \delta x, \delta x; t, \delta t) \, d(\delta x), \tag{A1.2}$$

where the integral is taken over all possible values of $\delta x$. In other words, the probability that the frequency of $A_1$ becomes $x$ at time $t + \delta t$ is the sum of the probabilities of the cases in which the gene frequency is $x - \delta x$ at time $t$ and then increases by $\delta x$ during the succeeding time interval of length $\delta t$. For a small $\delta t$, the probability distribution of $\delta x$ is effectively restricted to a small neighborhood around zero, so that the integrand in the right-hand side of (A1.2) may be expanded in a power series of $\delta x$:

$$\phi(p, x; t + \delta t) = \int \Big\{ \phi g - \delta x \frac{\partial}{\partial x} (\phi g) + \frac{(\delta x)^2}{2} \frac{\partial^2}{\partial x^2} (\phi g) - \cdots \Big\} \, d(\delta x), \tag{A1.3}$$

where $\phi = \phi(p, x; t)$ and $g = g(x, \delta x; t, \delta t)$. Neglecting higher-order terms involving $(\delta x)^3$ etc., noting that

$$\int g \, d(\delta x) = 1$$

and writing

$$M(x, t, \delta t) = \int (\delta x) g \, d(\delta x)$$

and

$$V(x, t, \delta t) = \int (\delta x)^2 g \, d(\delta x),$$

we obtain from (A1.3)

$$\phi(p, x; t+\delta t) = \phi(p, x; t) - \frac{\partial}{\partial x} \{M(x, t, \delta t)\phi(p, x; t)\} + \frac{1}{2}\frac{\partial^2}{\partial x^2} \{V(x, t, \delta t)\phi(p, x; t)\}.$$

Thus, at the limit of $\delta t \to 0$, noting

$$\frac{\partial \phi}{\partial t} = \lim_{\delta t \to 0} \frac{\phi(p, x; t+\delta t) - \phi(p, x; t)}{\delta t}$$

and

$$M_{\delta x} = \lim_{\delta t \to 0} \frac{M(x, t, \delta t)}{\delta t} = \lim_{\delta t \to 0} \frac{E_\delta(\delta x)}{\delta t},$$

$$V_{\delta x} = \lim_{\delta t \to 0} \frac{V(x, t, \delta t)}{\delta t} = \lim_{\delta t \to 0} \frac{E_\delta\{(\delta x)^2\}}{\delta t}, \qquad \text{(A1.4)}$$

we get (A1.1) as was to be shown.

It can also be shown that

$$F(x, t) = -\frac{1}{2}\frac{\partial}{\partial x} \{V_{\delta x}\phi(p, x; t)\} + M_{\delta x}\phi(p, x; t) \qquad \text{(A1.5)}$$

represents the net probability flux across the point $x$ at time $t$ (cf. Kimura 1964).

This may be used to derive Wright's formula (Wright 1938a) for the distribution of the gene frequency at equilibrium in which a stationary distribution is reached with respect to the frequency ($x$) of allele $A_1$ when the dispersing factor, such as random sampling of gametes, is balanced by the centripetal factor, such as reversible mutation.

If we denote by $\phi(x)$ the probability density of such an equilibrium distribution, then it is independent of the initial frequency $p$ and satisfies the relation

$$-\frac{1}{2}\frac{\partial}{\partial x}\{V_{\delta x}\phi(x)\}+M_{\delta x}\phi(p, x; t)=0, \qquad \text{(A1.6)}$$

because the net probability flux is zero at any point of $x$ in the interval (0, 1). Equation (A1.6) may immediately be integrated to give Wright's formula

$$\phi(x)=\frac{C}{V_{\delta x}}\exp\{2\int\frac{M_{\delta x}}{V_{\delta x}}\,dx\}, \qquad \text{(A1.7)}$$

where the constant $C$ is usually determined so that

$$\int_0^1 \phi(x)\,dx=1. \qquad \text{(A1.8)}$$

For example, in a population of effective size $N_e$, if there is reversible mutation between a pair of alleles such that $u$ is the mutation rate from $A_1$ to $A_2$ and $v$ is that in the reverse direction, then

$$M_{\delta x}=-ux+v(1-x)$$

and (A1.9)

$$V_{\delta x}=\frac{x(1-x)}{2N_e}.$$

Substituting these in (A1.7), we obtain

$$\phi(x)=Cx^{4N_e v-1}(1-x)^{4N_e u-1}, \qquad \text{(A1.10)}$$

where

$$C=\frac{\Gamma(4N_e(u+v))}{\Gamma(4N_e u)\Gamma(4N_e v)}. \qquad \text{(A1.11)}$$

Equation (A1.1) can be extended to cover cases involving two or more random variables. For example, if three alleles $A_1$, $A_2$ and $A_3$ are segregating in a locus, we let $\phi(p_1, p_2, x_1, x_2; t)$ be the probability density that the frequencies of $A_1$

and $A_2$ become $x_1$ and $x_2$ at time $t$, given that their frequencies are $p_1$ and $p_2$ at time 0, then

$$\frac{\partial \phi}{\partial t} = \frac{1}{2} \frac{\partial^2}{\partial x_1^2} (V_{\delta x_1} \phi) + \frac{1}{2} \frac{\partial^2}{\partial x_2^2} (V_{\delta x_2} \phi) + \frac{\partial^2}{\partial x_1 \partial x_2} (W_{\delta x_1 \delta x_2} \phi)$$
$$- \frac{\partial}{\partial x_1} (M_{\delta x_1} \phi) - \frac{\partial}{\partial x_2} (M_{\delta x_2} \phi), \qquad \text{(A1.12)}$$

where $W_{\delta x_1 \delta x_2}$ is the covariance between $\delta x_1$ and $\delta x_2$.

Returning to the case of two alleles at a locus, let $\phi(p, x; \tau, t)$ be the probability density that the frequency of $A_1$ becomes $x$ at time $t$, given that it is $p$ at time $\tau (\tau < t)$. So far, we have treated $x$ and $t$ as variables, and $p$ has been assumed fixed. Also, the initial time has been fixed and set to zero. We shall now look at the process retrospectively and regard $x$ and $t$ as fixed, while the initial frequency $p$ and initial time $\tau$ are variable. Then we have the Kolmogorov backward equation

$$-\frac{\partial \phi(p, x; \tau, t)}{\partial \tau} = \frac{V_{\delta p}}{2} \frac{\partial^2 \phi(p, x; \tau, t)}{\partial p^2}$$
$$+ M_{\delta p} \frac{\partial \phi(p, x; \tau, t)}{\partial p}. \qquad \text{(A1.13)}$$

This may be derived by noting that

$$\phi(p, x; \tau, t) = \int g(p, \delta p; \tau, \delta \tau) \phi(p + \delta p, x; \tau + \delta \tau, t) \, d(\delta p), \qquad \text{(A1.14)}$$

namely, the probability that the frequency of $A_1$ becomes $x$ at time $t$ given that it is $p$ at time $\tau$ is equal to the sum of the probabilities of cases in which the gene frequency changes first from $p$ to $p + \delta p$ during the time interval $(\tau, \tau + \delta \tau)$ and then changes from $p + \delta p$ to $x$ during the subsequent time interval $(\tau + \delta \tau, t)$. Expanding $\phi$ in the integral in terms of $\delta p$ and $\delta \tau$,

$$\phi(p + \delta p, x; \tau + \delta\tau, t) = \phi(p, x; \tau, t) + \delta p \frac{\partial}{\partial p} \phi(p, x; \tau, t)$$
$$+ \delta\tau \frac{\partial}{\partial \tau} \phi(p, x; \tau, t)$$
$$+ \frac{(\delta p)^2}{2} \frac{\partial^2}{\partial p^2} \phi(p, x; \tau, t) + \cdots.$$

Neglecting higher-order terms involving $(\delta\tau)^2$, $(\delta\tau)\delta p$, $(\delta p)^3$ etc., we obtain (A1.13) at the limit of $\delta t \to 0$, just as (A1.1) was derived from (A1.2).

When the process is time-homogeneous, i.e. when $g(p, \delta p; \tau, \delta\tau)$ is independent of the third argument $\tau$ so that $M_{\delta p}$ and $V_{\delta p}$ are independent of time, (A1.13) reduces to

$$\frac{\partial \phi}{\partial t} = \frac{V_{\delta p}}{2} \frac{\partial^2 \phi}{\partial p^2} + M_{\delta p} \frac{\partial \phi}{\partial p}, \tag{A1.15}$$

where $\phi$ now denotes $\phi(p, x; t)$ and has the same meaning as $\phi$ in (A1.1), namely, it is the probability density that the gene frequency becomes $x$ at time $t$ given that it is $p$ at time zero.

One of the very important applications of the backward equation is its use in deriving the probability of gene fixation in a finite population, as may be seen by noting that gene fixation corresponds to $x = 1$. In fact, equation (A1.15) was first introduced in population genetics to treat this problem (Kimura 1957, 1962). A more general equation (A1.13), or a form equivalent to it, can be used to obtain the probability of fixation of a mutant gene when its selective advantage decreases with time (Kimura and Ohta 1970b). We note here that if we denote by $u(p, \tau, t)$ the probability that $A_1$ becomes fixed (established) in the population *by* time $t$, given that it is $p$ at time $\tau (\tau < t)$, then we can replace $u$ for $\phi$ in equations (A1.13) and (A1.14).

In the special case of time homogeneity, let $u(p, t)$ be the probability that $A_1$ becomes fixed in the population by the time $t$, given that it is $p$ at the start. Then we have

$$\frac{\partial u(p, t)}{\partial t} = \frac{V_{\delta p}}{2} \frac{\partial^2 u(p, t)}{\partial p^2} + M_{\delta p} \frac{\partial u(p, t)}{\partial p}. \qquad \text{(A1.16)}$$

The required probability may then be obtained by solving this equation with the boundary conditions

$$u(0, t) = 0, \qquad u(1, t) = 1, \qquad \text{(A1.17)}$$

namely, the probability is always zero if the initial gene frequency is zero, and it is unity if the gene is already fixed at the start.

Of particular interest is the probability of ultimate fixation denoted by

$$u(p) = \lim_{t \to \infty} u(p, t).$$

Since $\partial u / \partial t = 0$ for this probability, (A1.16) and (A1.17) are reduced to

$$\frac{V_{\delta p}}{2} \frac{\mathrm{d}^2 u(p)}{\mathrm{d}p^2} + M_{\delta p} \frac{\mathrm{d}u(p)}{\mathrm{d}p} = 0 \qquad \text{(A1.18)}$$

and

$$u(0) = 0, \qquad u(1) = 1. \qquad \text{(A1.19)}$$

The solution is then given by

$$u(p) = \frac{\int_0^p G(x)\,\mathrm{d}x}{\int_0^1 G(x)\,\mathrm{d}x}, \qquad \text{(A1.20)}$$

where

$$G(x) = \exp\{-2 \int \frac{M_{\delta x}}{V_{\delta x}}\,\mathrm{d}x\} \qquad \text{(A1.21)}$$

(Kimura 1962).

For example, consider a random mating population and assume that the mutant gene ($A_1$) has selective advantage $s$ in the homozygote and $sh$ in the heterozygote over its allele ($A_2$) so that

$$M_{\delta p} = sp(1-p)\{h + (1-2h)p\}. \qquad \text{(A1.22)}$$

Assume also that $N_e$ is the "variance" effective number so that

$$V_{\delta p} = \frac{p(1-p)}{2N_e}. \qquad \text{(A1.23)}$$

Then, from (A1.21), we have

$$G(x) = -2N_e s\{(2h-1)x(1-x) + x\}.$$

Substituting this in (A1.20), we obtain

$$u(p) = \frac{\int_0^p e^{-2S\{(2h-1)x(1-x)+x\}}\,\mathrm{d}x}{\int_0^1 e^{-2S\{(2h-1)x(1-x)+x\}}\,\mathrm{d}x} \qquad \text{(A1.24)}$$

where $S = N_e s$.

Another interesting problem relating to gene fixation is how long a mutant gene will take until it reaches fixation, excluding the cases of its eventual loss. In particular, the average number of generations until fixation is important in considering the rate of evolution by gene substitution (Kimura and Ohta 1969a).

Let

$$T_1(p) = \int_0^\infty t\,\frac{\partial u(p,t)}{\partial t}\,\mathrm{d}t, \qquad \text{(A1.25)}$$

then

$$\bar{t}_1(p) = \frac{T_1(p)}{u(p)} \qquad \text{(A1.26)}$$

represents the average number of generations until fixation excluding the cases of eventual loss.

Since $u(p)$ is already given by (A1.20), the problem is solved if we can obtain an explicit expression for $T_1(p)$. To this end, we differentiate each term of (A1.16) with respect to $t$, multiply each of the resulting terms by $t$ and then integrate them with respect to $t$ from 0 to $\infty$. This yields

$$\int_0^\infty t\frac{\partial^2 u(p,t)}{\partial t^2}\,\mathrm{d}t = \frac{V_{\delta p}}{2}\frac{\partial^2}{\partial p^2}T_1(p) + M_{\delta p}\frac{\partial}{\partial p}T_1(p),$$

but the left-hand term reduces to $-u(p)$ if we assume that $t\partial u(p,t)/\partial t$ vanishes at $t=\infty$, because

$$\int_0^\infty t\frac{\partial^2 u(p,t)}{\partial t^2}\,\mathrm{d}t = \left[t\frac{\partial u(p,t)}{\partial t}\right]_0^\infty - \int_0^\infty \frac{\partial u(p,t)}{\partial t}\,\mathrm{d}t$$

$$= -u(p,\infty) = -u(p).$$

Therefore, we have the ordinary differential equation,

$$T_1''(p) + \frac{2M_{\delta p}}{V_{\delta p}}T_1'(p) + \frac{2u(p)}{V_{\delta p}} = 0. \qquad \text{(A1.27)}$$

The appropriate boundary conditions are

$$\lim_{p\to 0}\bar{t}_1(p) = \text{finite}$$

and

$$\bar{t}_1(1) = 0,$$

and if we solve (A1.27) under these conditions, we get

$$T_1(p) = u(p)\int_p^1 \psi(\xi)u(\xi)\{1-u(\xi)\}\,\mathrm{d}\xi$$

$$+\{1-u(p)\}\int_0^p \psi(\xi)u^2(\xi)\,\mathrm{d}\xi, \qquad \text{(A1.28)}$$

where

$$\psi(x) = \frac{2 \int_0^1 G(\xi)\, d\xi}{V_{\delta x} G(x)}. \qquad \text{(A1.29)}$$

The average number of generations until fixation of a mutant gene ($A_1$) with initial frequency $p$ is then given by (A1.26) with $T_1(p)$ given by (A1.28) and $u(p)$ by (A1.20).

The backward equation (A1.15) can also be used to obtain the frequency distribution of mutants under a steady flux of mutations (Kimura 1969b). Let us consider a situation in which a very large (practically infinite) number of sites are available for mutation. More specifically, we assume that the total number of sites per individual is so large and the mutation rate per site is so low that whenever a mutant appears, it represents a mutation at a different and previously homallelic site, i.e. a site in which no mutant forms are currently segregating in the population. In this model, "site" refers to a single nucleotide site rather than a conventional genetic locus, but the following treatment is also appropriate to a small group of nucleotides such as a codon.

Let $\nu_m$ be the number of mutants that appear each generation in the entire population each representing a mutation at a different site in the genome. If we consider a particular site in which a mutant has appeared, and if we denote by $p$ the initial frequency of the mutant in the population, then the probability density, $\phi(p, x; t)$, that its frequency becomes $x$ after $t$ generations satisfies equation (A1.15).

Since a mutant that appears in a finite population is either lost from the population or fixed in it within a finite length of time, if the production of mutants continues at a constant rate over a long period, a steady state will be reached with respect to frequency distribution of mutants among different sites. Note that the steady state distribution here is concerned with only those sites in which mutants are neither fixed nor lost, i.e. we consider only those sites in which mutants are segregating in the population. We shall denote by $\phi(p, x)$ the steady state distribution such that $\phi(p, x)\,dx$ is the expected number (not relative frequency) of sites in which the frequencies of mutants are in the range $x \sim x + dx$. Since new mutants appear in $\nu_m$ sites in each generation, the contribution of the mutants that appeared $t$ generations earlier to the present frequency class $x$ is $\nu_m \phi(p, x; t)$. Thus, considering all the contributions made by mutations in the past, we have

$$\Phi(p, x) = \nu_m \int_0^\infty \phi(p, x; t)\,dt. \qquad \text{(A1.30)}$$

Let $I_f(p)$ be the expectation (functional) of an arbitrary function $f(x)$ with respect to this distribution, i.e.

$$I_f(p) = \int_0^1 f(x)\Phi(p, x)\,dx. \qquad \text{(A1.31)}$$

Note that the integral is over the open interval (0, 1), since it is a continuous approximation to the sum over the discrete intervals from $x = 1/(2N)$ to $1 - 1/(2N)$.

Multiplying each term of (A1.15) by $\nu_m f(x)$ and then integrating the resulting terms first with respect to $x$ over the interval (0, 1) and then with respect to $t$ over $(0, \infty)$, we get

$$\int_0^\infty \frac{\partial}{\partial t}\{\nu_m \int_0^1 f(x)\phi(p, x; t)\,dx\}\,dt$$

$$= \frac{V_{\delta p}}{2}\frac{\partial^2}{\partial p^2} I_f(p) + M_{\delta p}\frac{\partial}{\partial p} I_f(p). \qquad \text{(A1.32)}$$

If we note the conditions,

$$\phi(p, x; \infty) = 0 \qquad \text{for} \qquad 0 < x < 1,$$

and (A1.33)

$$\phi(p, x; 0) = \delta(x - p),$$

in which $\delta(\cdot)$ is the Dirac delta function,* the left-hand side of (A1.32) is reduced to a simple form as follows,

$$\left[\nu_m \int_0^1 f(x)\phi(p, x; t)\, \mathrm{d}x\right]_{t=0}^{t=\infty} = -\nu_m \int_0^1 f(x)\delta(x - p)\, \mathrm{d}x = -\nu_m f(p).$$

Of the two conditions given in (A1.33), the first follows from the fact that each mutant becomes either fixed or lost within a finite length of time, and the second represents the fact that each has the initial frequency $p$.

Thus, (A1.32) yields the ordinary differential equation

$$\tfrac{1}{2}V_{\delta p} I_f''(p) + M_{\delta p} I_f'(p) + \nu_m f(p) = 0. \qquad \text{(A1.34)}$$

Although $M_{\delta p}$ and $V_{\delta p}$ as functions of $p$ are rather general, in a typical situation we assume that they are given by (A1.22) and (A1.23).

The solution of this equation which satisfies the boundary conditions

$$I_f(0) = I_f(1) = 0 \qquad \text{(A1.35)}$$

may be expressed in the form (Kimura 1969b):

$$I_f(p) = \{1 - u(p)\} \int_0^p \psi_f(\xi) u(\xi)\, \mathrm{d}\xi + u(p) \int_p^1 \psi_f(\xi)\{1 - u(\xi)\}\, \mathrm{d}\xi, \qquad \text{(A1.36)}$$

where $u(p)$ is the probability of ultimate fixation given by (A1.20), and

* The Dirac delta function $\delta(x)$ is a special sort of function which is zero for $x \neq 0$ and becomes indefinitely large for $x = 0$. An important property of this function is that for any $f(x)$ we have

$$\int f(x)\delta(x - y)\, \mathrm{d}x = f(y)$$

if the range of integration contains the point $x = 0$.

$$\psi_f(\xi) = \frac{2v_m f(\xi) \int_0^1 G(x)\, dx}{V_{\delta\xi} G(\xi)}$$

$$= \frac{2v_m f(\xi)}{V_{\delta\xi} u'(\xi)} \quad \text{(A1.37)}$$

in which $u'(\xi) = du(\xi)/d\xi$. Boundary conditions (A1.35) correspond to the fact that "mutations" with $p = 0$ and $p = 1$ do not contribute to the distribution at the segregating sites.

Formula (A1.36) should have wide applicability since various statistics relating to the steady flux distribution $\Phi(p, x)$ can be obtained by assigning various functions of $x$ to $f$ in $I_f(p)$. For example, $I_f(p)$ with $f = 2x(1 - x)$ gives the mean number of heterozygous sites per individual, $I_f(p)$ with $f = 1$ gives the total number of segregating sites in the population at any given moment, $I_f(p)$ with $f = s - \{sx^2 + sh2x(1 - x)\}$ gives the substitutional load in a population of effective size $N_e$ if $M_{\delta p}$ and $V_{\delta p}$ is given by (A1.22) and (A1.23), and so on. Furthermore, by putting $f(x) = \delta(x - y)$ in $I_f(p)$ in which $\delta(\cdot)$ is the Dirac delta function, we can derive the distribution function itself, for

$$I_f(p) = \int_0^1 \delta(x - y)\Phi(p, x)\, dx = \Phi(p, y)$$

with this assignment. This leads to

$$\Phi(p, y) = \frac{2v_m u(p)\{1 - u(y)\}}{V_{\delta y} u'(y)} \quad \text{(A1.38)}$$

for $p \leqslant y < 1$, and

$$\Phi(p, y) = \frac{2v_m\{1 - u(p)\}u(y)}{V_{\delta y} u'(y)} \quad \text{(A1.38a)}$$

for $0 < y \leqslant p$.

Although $p$ may take any value between 0 and 1 exclusive, a typical value is $p = 1/(2N)$, in which $N$ is the actual population number, because, with very low mutation rate per site, each mutant is likely to be represented only once at the moment of its appearance. Then, only formula (A1.38) is needed to describe the distribution. In this case, writing $\Phi(x)$ for $\Phi(p, x)$, (A1.38) reduces to approximately

$$\Phi(x) = \frac{2v}{V_{\delta x} G(x)} \frac{\int_x^1 G(x)\,dx}{\int_0^1 G(x)\,dx}, \tag{A1.39}$$

where $v = \nu_m/(2N)$ is the mutation rate per gamete per generation and $G(x)$ is given by (A1.21). The formula is an approximation to the steady flux distribution such that $\Phi(x)\,dx$ gives the number of sites in which the mutant frequency is $x$ in the interval $1/(2N) \leqslant x \leqslant 1 - 1/(2N)$.

Returning to formula (A1.36), we note that if the mutants are selectively neutral, $M_{\delta x} = 0$ and therefore $u(p) = p$, $G(x) = 1$, $\psi_f(\xi) = 2\nu_m f(\xi)/V_{\delta\xi}$ in the formula. Then, the number of heterozygous sites per individual, $H(p)$, in a population of effective size $N_e$ may then be obtained by putting $f(x) = 2x(1 - x)$ and $V_{\delta x} = x(1 - x)/(2N_e)$. This yields

$$H(p) = 4N_e \nu_m p(1 - p). \tag{A1.40}$$

## A2. *Number of Neutral Alleles in a Finite Population*

Consider a particular locus and assume that there are $K$ possible allelic states $A_1, A_2, \ldots A_K$. Let $u$ be the mutation rate per locus per generation and assume that the mutation rate is equal to all alleles so that each allele mutates to one of the remaining $(K - 1)$ alleles with rate $u/(K - 1)$. In a random mating population having "variance" effective number $N_e$, if there is no selection, the mean and the variance of the change per generation of the frequency of a particular allele, say $A_1$, are respectively

$$M_{\delta x} = -ux + v_1(1 - x)$$

and (A2.1)

$$V_{\delta x} = \frac{x(1 - x)}{2N_e},$$

where $v_1 = u/(K - 1)$, and $x$ stands for the frequency of $A_1$. Throughout this section, to simplify expressions, the letter $x$ rather than $x_1$ (or more generally $x_i$) will be used to represent the frequency of a particular allele, unless simultaneous consideration of all the allelic frequencies is required.

The set of $M_{\delta x}$ and $V_{\delta x}$ in (A2.1) have the same form as those in (A1.9) in the previous section, except that symbol $v_1$ is now used rather than $v$. In other words, in the present model of neutral isoalleles with equal mutation rates, the problem of considering the frequency distribution of a particular allele among $K$ alleles is reduced to that of the case of two alleles by designating the remaining $(K - 1)$ alleles collectively as $A_2$.

Thus, the frequency distribution of $x$ at equilibrium is given by (A1.10) and (A1.11) if $v$ is replaced by $v_1$. Namely,

$$\phi(x) = \frac{\Gamma(A + B)}{\Gamma(A)\Gamma(B)} x^{A-1}(1 - x)^{B-1}, \qquad \text{(A2.2)}$$

where $A = 4N_e u/(K - 1)$ and $B = 4N_e u$.

The expected frequency of the $A_1A_1$ homozygote is

$$E(x^2) = \int_0^1 x^2\phi(x)\, dx = \frac{(A + 1)}{K(A + B + 1)}, \qquad \text{(A2.3)}$$

and therefore, considering all the $K$ alleles, the expected total frequency of homozygotes is

$$f = E\left(\sum_1^K x_i^2\right) = \frac{A + 1}{A + B + 1}. \qquad \text{(A2.4)}$$

The frequency of heterozygotes is

$$H = 1 - f = \frac{B}{A + B + 1}. \tag{A2.5}$$

The "effective number" of alleles maintained in the population is defined as the reciprocal of the sum of the squares of the allelic frequencies, or $1/f$, so

$$n_e = \frac{1}{f} = \frac{A + B + 1}{A + 1}. \tag{A2.6}$$

This is usually much smaller than the "average number" of alleles, $n_a$, defined as the reciprocal of the average frequency of alleles currently segregating within a single population. Since the probability $P_1(0)$ that a particular allele is temporarily lost from the population is given by

$$P_1(0) = \int_0^{1/(2N)} \phi(x)\, dx, \tag{A2.7}$$

in which $N$ is the actual population number, the average frequency of this allele in the population, given that it is represented at least once in the population, is

$$\bar{x}[x \neq 0] = \frac{E(x)}{1 - P_1(0)} = \frac{1/K}{1 - P_1(0)}. \tag{A2.8}$$

Thus, the average number of alleles

$$\begin{aligned} n_a &= K\{1 - P_1(0)\} \\ &= \frac{K\Gamma(A + B)}{\Gamma(A)\Gamma(B)} \int_{1/(2N)}^{1} x^{A-1}(1 - x)^{B-1}\, dx, \end{aligned} \tag{A2.9}$$

in which $A = B/(K - 1)$ and $B = 4N_e u$.

In observational studies of natural populations, very rare alleles are unlikely to be detected, so we shall call a population "monomorphic" if the sum of the frequencies of "variant" alleles that happen to exist in the population is $q$ or less. The value of $q$ is arbitrary, but a reasonable value is $q = 0.01$. With this definition, the probability that a population is monomorphic is

$$P_{\text{mono}} = K \int_{1-q}^{1} \phi(x)\,\mathrm{d}x. \qquad \text{(A2.10)}$$

If the number of allelic states is so large that each new mutant represents an allelic state not preexisting in the population, we may put $K \to \infty$ in the above formulae, yielding the following formulae:

the effective number of alleles

$$n_e = 4N_e u + 1, \qquad \text{(A2.11)}$$

the average number of alleles

$$n_a = 4N_e u \int_{1/(2N)}^{1} (1-x)^{4N_e u - 1} x^{-1}\,\mathrm{d}x, \qquad \text{(A2.12)}$$

the frequency of heterozygotes

$$H = \frac{4N_e u}{4N_e u + 1}, \qquad \text{(A2.13)}$$

the probability of a population being monomorphic

$$P_{\text{mono}} = q^{4N_e u}. \qquad \text{(A2.14)}$$

For detailed discussions of the number of neutral isoalleles in a finite population, readers may refer to Kimura and Crow (1964) and Kimura (1968b).

### A3. *Basic Equations for Deriving the Moments of Distribution*

The method in this section has been developed by Ohta and Kimura (1969b, 1971a) to treat the problem of linkage disequilibrium in a finite population based on diffusion models. Although the applicability of this method is mainly restricted to the cases in which the mean rates of change of gene frequencies are linear or linearized with good approximation, it has proved quite useful in the problem of linkage disequilibrium between two loci for which simultaneous treatment of three independent random variables is usually required.

Consider the single variable case first. Let $x_t$ be the frequency of a particular allele at time $t$, and let $\delta x_t$ be an increment of $x_t$ during the subsequent short time interval of length $\delta t$.

$$x_{t+\delta t} = x_t + \delta x_t. \qquad \text{(A3.1)}$$

Let $f(x)$ be a polynomial in $x$ and consider the expected value of this function $E\{f(x)\}$. Then, from (A3.1),

$$E\{f(x_{t+\delta t})\} = E_\phi E_\delta\{f(x_t + \delta x_t)\}, \qquad \text{(A3.2)}$$

where $E_\phi$ is the operator for taking the expectation with respect to the probability distribution $\phi$ at time $t$, and $E_\delta$ is the operator for taking the expectation with respect to the change $\delta x_t$. Expanding $f(x_t + \delta x_t)$ in the right-hand side of (A3.2) in terms of $\delta x_t$, and neglecting higher-order terms involving $(\delta x_t)^3$ etc., we have

$$E_\phi E_\delta\{f(x_t + \delta x_t)\} =$$

$$E_\phi\{f(x_t) + E_\delta(\delta x_t)f'(x_t) + 1/2\ E_\delta(\delta x_t)^2 f''(x_t)\}.$$

Thus, (A3.2) yields

$$\frac{E\{f(x_{t+\delta t})\} - E\{f(x_t)\}}{\delta t} = E\left\{\frac{E_\delta(\delta x_t)}{\delta t} f'(x_t) + \frac{1}{2}\frac{E_\delta(\delta x_t)^2}{\delta t} f''(x_t)\right\}.$$

At the limit of $\delta t \to 0$, noting (A1.4) in section A1, we obtain the basic equation

$$\frac{\mathrm{d}}{\mathrm{d}t} E\{f(x)\} = E\left\{\frac{V_{\delta x}}{2} f''(x) + M_{\delta x} f'(x)\right\}, \qquad \text{(A3.3)}$$

where $M_{\delta x}$ and $V_{\delta x}$ are the mean and the variance of the change in gene frequency per generation, and $E$ now denotes $E_\phi$, i.e.

$$E\{f(x)\} = \int_0^1 f(x)\phi(p, x; t)\,dx, \qquad \text{(A3.4)}$$

in which $\phi(p, x; t)$ is the probability density that the gene frequency becomes $x$ at time $t$ given that it is $p$ at time 0.

For example, in the case of random genetic drift due to random sampling of gametes, $M_{\delta x} = 0$, $V_{\delta x} = x(1-x)/(2N_e)$ in (A3.3). Let $f(x) = x$, then $f'(x) = 1$, $f''(x) = 0$ and (A3.3) becomes

$$\frac{d}{dt} E(x) = 0,$$

from which we obtain

$$E(x) = \bar{x} = \text{constant}, \qquad \text{(A3.5)}$$

which we shall denote by $p$. Next, let $f(x) = (x - \bar{x})^2 = (x - p)^2$, then (A3.3) becomes

$$\frac{d}{dt} V_t = E\left\{\frac{x(1-x)}{2N_e}\right\} = \frac{1}{2N_e}\{p(1-p) - V_t\},$$

where $V_t = E\{(x-p)^2\}$ is the variance of gene frequency at time $t$. Solving this equation, we obtain

$$V_t = p(1-p)[1 - e^{-t/2N_e}]. \qquad \text{(A3.6)}$$

Returning to the basic equation (A3.3), it may readily be seen that in the stationary state in which the probability flux is zero at every point in the interval (0, 1), the left-hand side of (A3.3) becomes zero and we have

$$E\left\{\frac{V_{\delta x}}{2} f''(x) + M_{\delta x} f'(x)\right\} = 0. \qquad \text{(A3.7)}$$

The distribution function $\phi$ in this case is independent of both $p$ and $t$, and is given by Wright's formula (A1.7). For example, in the stationary state attained under reversible mutation between a pair of alleles, with $M_{\delta x}$ and $V_{\delta x}$ given by (A1.9), if we let $f(x) = x$, then $f'(x) = 1$ and $f''(x) = 0$ in (A3.7), and we have

$$E\{-ux + v(1 - x)\} = 0,$$

or

$$E(x) = \bar{x} = v/(u + v). \tag{A3.8}$$

Next, letting $f(x) = x^2$ so that $f'(x) = 2x, f''(x) = 2$ in (A3.7), we get

$$E\left\{\frac{x(1-x)}{2N_e} + (-ux + v(1-x))2x\right\} = 0,$$

from which we obtain

$$E\{(x - \bar{x})^2\} = \sigma_x^{\,2} = \frac{\bar{x}(1-\bar{x})}{1 + 4N_e(u + v)}. \tag{A3.9}$$

The stationary state equation (A3.7) does not hold for the equilibrium state attained under steady flux of mutations. In order to derive an equation applicable to such a case, let us consider a situation in which mutation occurs at $x = p$ each generation and mutants go into irreversible fixation into either class $x = 0$ or $x = 1$; each mutant allele is either lost from the population or fixed in it within a finite length of time. This is the situation treated in the latter part of section A1, with distribution given by $\Phi(p, x)$ in equation (A1.30). The terms $M_{\delta x}$ and $V_{\delta x}$ in that treatment do not contain the effect of mutation, but the mutants appears at $\nu_m$ sites each generation in the entire population. So, writing $\Delta_{\text{mut}}E(f)$ to denote the input with respect to $E(f)$ by mutation per unit time, we have

$$\Delta_{\text{mut}}E(f) + \frac{\mathrm{d}}{\mathrm{d}t}\{Ef(x)\} = 0, \tag{A3.10}$$

at the state in which the input by mutation is balanced by

the decay of variability by random drift. Thus, we obtain the steady flux equation

$$E\left\{\frac{V_{\delta x}}{2} f''(x) + M_{\delta x} f'(x)\right\} = -\Delta_{\text{mut}} E(f). \qquad \text{(A3.11)}$$

It is important to note that this equation is valid only for $f(x)$ which vanishes both at $x = 0$ and $x = 1$. More precisely, $f(x)$ must be such that $f(x)\Phi(p, x)$ becomes zero both at $x = 0$ and $x = 1$, because the steady state distribution in this case refers only to unfixed classes ($1/2N \leqq x \leqq 1 - 1/2N$), and the newly fixed terminal classes ($x = 0$ and $x = 1$) must not contribute to the left-hand side of equation (A3.11).

As an example, let us assume that mutation is selectively neutral and that in each generation new mutations occur in the entire population in $\nu_m$ sites. Let $f(x) = 2x(1 - x)$, then $E\{f(x)\} \equiv H(p)$, in which $p$ is the initial frequency of each mutant, represents the number of heterozygous sites per individual. Since $M_{\delta x} = 0$, $V_{\delta x} = x(1 - x)/(2N_e)$, $f''(x) = -4$ and $\Delta_{\text{mut}}E(f) = \nu_m 2p(1 - p)$ in this example, (A3.11) yields

$$-\frac{1}{2N_e} E\{2x(1 - x)\} = -\nu_m 2p(1 - p)$$

or

$$H(p) = 4N_e \nu_m p(1 - p)$$

which agrees with (A1.40) obtained by a different method.

So far, we have considered the single variable case, but the real use of the present method lies in its application to multi-variate cases, for which the method achieves a great simplification.

Suppose that there are $n$ independent random variables $x_1, x_2, \ldots, x_n$. Then, it can readily be shown by using a procedure similar to that used to derive equation (A3.3), that the basic equation for the moments becomes

$$\frac{\mathrm{d}}{\mathrm{d}t} E\{f(x_1, x_2, \ldots, x_n)\} = E\{Lf(x_1, x_2, \ldots, x_n)\}, \tag{A3.12}$$

where $L$ stands for the differential operator such that

$$L = \frac{1}{2} \sum_{i=1}^{n} V_{\delta x_i} \frac{\partial^2}{\partial x_i^2} + \sum_{i>j} W_{\delta x_i \delta x_j} \frac{\partial^2}{\partial x_i \partial x_j} + \sum_{i=1}^{n} M_{\delta x_i} \frac{\partial}{\partial x_i}, \tag{A3.13}$$

in which $M$, $V$ and $W$ respectively denote the mean, variance and covariance with respect to changes appearing as subscripts.

Thus, we have

$$E\{Lf(x_1, x_2, \ldots, x_n)\} = 0 \tag{A3.14}$$

for the stationary state, and

$$E\{Lf(x_1, x_2, \ldots, x_n)\} = -\Delta_{\text{mut}} E(f) \tag{A3.15}$$

for the steady flux state.

As an application of equation (A3.14), consider two linked loci $A$ and $B$. We assume that at locus $A$ a pair of overdominant alleles $A_1$ and $A_2$ are segregating with respective frequencies $p$ and $1-p$, and at locus $B$ a pair of neutral alleles $B_1$ and $B_2$ are segregating with respective frequencies $q$ and $1-q$. Let $D = X_1X_4 - X_2X_3$ be the index of linkage disequilibrium in which $X_1$, $X_2$, $X_3$ and $X_4$ are respectively the frequencies of the four chromosomes $A_1B_1$, $A_1B_2$, $A_2B_1$ and $A_2B_2$. We shall denote by $c$ the recombination fraction between the two loci, by $u$ the mutation rate from $B_1$ to $B_2$ and by $v$ that of the reverse direction. If we assume that overdominance at $A$ locus is so strong that the frequency $p$ remains practically constant, then we have two random variables $q$ and $D$. Thus (A3.14) becomes

$$E\left\{\frac{V_{\delta q}}{2}\frac{\partial^2 f}{\partial q^2} + W_{\delta q \delta D}\frac{\partial^2 f}{\partial q \partial D} + \frac{V_{\delta D}}{2}\frac{\partial^2 f}{\partial D^2} + M_{\delta q}\frac{\partial f}{\partial q} + M_{\delta D}\frac{\partial f}{\partial D}\right\} = 0. \quad \text{(A3.16)}$$

Let $N_e$ be the variance effective number of the population. Then, after some computation, we find that the variances, covariance and means in this equation are given as follows:

$$\begin{aligned}
V_{\delta q} &= \{q(1-q) - D^2/p(1-p)\}/2N_e \\
W_{\delta q \delta D} &= \{(1-2q)D - (2p-1)D^2/p(1-p)\}/2N_e \\
V_{\delta D} &= \{p(1-p)q(1-q) + (1-2p)(1-2q)D \\
&\quad - [1 - 3p(1-p)D^2]/p(1-p)\}/2N_e \qquad \text{(A3.17)} \\
M_{\delta q} &= v - (u+v)q \\
M_{\delta D} &= -(c+u+v)D.
\end{aligned}$$

Letting $f = D$ in (A3.16), we get $E(D) = 0$. Also, letting $f = q$, we get $E(q) = \bar{q} = v/(u+v)$. Furthermore, by letting successively $f = D^2$, $f = q^2$ and $f = qD$ in (A3.16), we obtain a set of equations,

$$\begin{aligned}
&p(1-p)X + 2(1-2p)Y + [p^{-1}(1-p)^{-1} - 3 \\
&\qquad + 4N_e(c+u+v)]Z = p(1-p)\bar{q}, \\
&[1 + 4N_e(u+v)]X + [p(1-p)]^{-1}Z = (1+4N_e v)\bar{q} \qquad \text{(A3.18)}
\end{aligned}$$

and

$$[2 + 2N_e(c+2u+2v)]Y - (2p-1)[p(1-p)]^{-1}Z = 0,$$

where $X = E(q^2)$, $Y = E(qD)$ and $Z = E(D^2)$. By solving these simultaneous equations for $X$, $Y$ and $Z$, we obtain the formula for the standard linkage deviation $\sigma_d^2 \equiv E(D^2)/E\{p(1-p)q(1-q)\}$.

$$\sigma_d^2 = 1 \Big/ \left\{ 1 + 4N_e(c + u + v) + \frac{(1 - 2p)^2}{p(1 - p)} \frac{N_e(c + 2u + 2v)}{1 + N_e(c + 2u + 2v)} \right\}. \tag{A3.19}$$

### A4. *Maintenance of Overdominant Alleles in a Partially Self-fertilizing Population*

Let $A_1$ and $A_2$ be a pair of overdominant alleles and let $P_{11}$, $2P_{12}$ and $P_{22}$ be the frequencies of the three genotypes $A_1A_1$, $A_1A_2$ and $A_2A_2$ in a population before selection ($P_{11} + 2P_{12} + P_{22} = 1$). We assume that the population is so large that random fluctuation of gene frequencies may be neglected. We consider a model of discrete generation time and assume that in each generation a fraction $R$ of individuals is produced by random mating and fraction $S$ by self-fertilization ($R + S = 1$).

Let $1 - s_1$, 1 and $1 - s_2$ be respectively the relative selective values of $A_1A_1$, $A_1A_2$ and $A_2A_2$. In other words, $s_1$ and $s_2$ are selection coefficients against the two homozygotes ($1 \geqslant s_1 > 0$, $1 \geqslant s_2 > 0$).

Then, if $x_1'$ and $x_2'$ are the frequencies of $A_1$ and $A_2$ among gametes produced by individuals of the present generation ($x_1' + x_2' = 1$), the zygotic frequencies in the next generation before selection are

$$\begin{aligned} P_{11}' &= Rx_1'^2 + S\{(1 - s_1)P_{11} + (1/2)P_{12}\}/\bar{w} \\ 2P_{12}' &= 2Rx_1'x_2' + SP_{12}/\bar{w} \\ P_{22}' &= R(x_2')^2 + S\{(1 - s_2)P_{22} + (1/2)P_{12}\}/\bar{w} \end{aligned} \tag{A4.1}$$

where $\bar{w}$ is the mean selective value of the population

$$\bar{w} = 1 - s_1P_{11} - s_2P_{22}, \tag{A4.2}$$

and

$$\begin{aligned} x_1' &= \{(1 - s_1)P_{11} + P_{12}\}/\bar{w} \\ x_2' &= \{(1 - s_2)P_{22} + P_{12}\}/\bar{w}. \end{aligned} \qquad \text{(A4.3)}$$

It may readily be seen by adding $P_{11}'$ and $P_{12}'$ in (A4.1) that $x_1'$ is the frequency of $A_1$ in the next generation before selection ($x_1' = P_{11}' + P_{12}'$), and similarly, $x_2'$ is the corresponding frequency of $A_2$. We shall denote by $x_1$ and $x_2$ the frequencies of $A_1$ and $A_2$ before selection in the present generation, i.e.

$$\begin{aligned} x_1 &= P_{11} + P_{12}, \\ x_2 &= P_{12} + P_{22}. \end{aligned} \qquad \text{(A4.4)}$$

At equilibrium, allelic frequencies remain constant from generation to generation so that

$$\frac{x_1}{x_2} = \frac{x_1'}{x_2'} = \frac{(1 - s_1)P_{11} + P_{12}}{(1 - s_2)P_{22} + P_{12}} = \frac{x_1 - s_1P_{11}}{x_2 - s_2P_{22}}.$$

This gives

$$s_1x_2P_{11} = s_2x_1P_{22}. \qquad \text{(A4.5)}$$

Also, at equilibrium, $P_{12}' = P_{12}$, $x_1' = x_1$, $x_2' = x_2$ and we have, from (A4.1),

$$2P_{12} = 2Rx_1x_2 + SP_{12}/\bar{w}. \qquad \text{(A4.6)}$$

In order to solve the set of equations (A4.5) and (A4.6), let

$$\begin{aligned} P_{11} &= (1 - \hat{f})x_1^2 + \hat{f}x_1 \\ P_{12} &= (1 - \hat{f})x_1x_2 \\ P_{22} &= (1 - \hat{f})x_2^2 + \hat{f}x_2 \end{aligned} \qquad \text{(A4.7)}$$

where $\hat{f}$ is formally equivalent to the inbreeding coefficient.

Substituting (A4.7) in (A4.5), we get the gene frequency

$$\hat{x}_1 = \frac{s_2 - s_1\hat{f}}{(s_1 + s_2)(1 - \hat{f})} \qquad (\hat{f} \neq 1), \qquad \text{(A4.8)}$$

and homozygote frequencies

$$\hat{P}_{11} = \hat{x}_1 i_s(1 + \hat{f})/s_1$$
$$\hat{P}_{22} = \hat{x}_2 i_s(1 + \hat{f})/s_2 \qquad \text{(A4.9)}$$

at equilibrium, where

$$i_s = \frac{s_1 s_2}{s_1 + s_2}. \qquad \text{(A4.10)}$$

Also, substituting (A4.7) in (A4.6), we obtain

$$2(1 - \hat{f}) = 2R + S(1 - \hat{f})/\hat{\bar{w}}, \qquad \text{(A4.11)}$$

where

$$\hat{\bar{w}} = 1 - i_s(1 + \hat{f})$$

from (A4.2) using (A4.9).

The equation determining $\hat{f}$ may then be derived from (A4.11):

$$2i_s\hat{f}^2 - (2 - S - 2i_sR)\hat{f} + S(1 - 2i_s) = 0. \qquad \text{(A4.12)}$$

This has two positive roots, but only one of them, i.e.

$$\hat{f} = \{2(1 - i_s) - (1 - 2i_s)S - \sqrt{D}\}/(4i_s) \qquad \text{(A4.13)}$$

is pertinent, where

$$D = 4(1 - i_s)^2 + (1 - 2i_s)^2S^2 - 4(1 + i_s)(1 - 2i_s)S. \qquad \text{(A4.14)}$$

The other root is irrelevant, since it does not lead to $\hat{f} = 0$ at $S = 0$.

In order that the gene frequency ($\hat{x}_1$) given by (A4.8) lie between 0 and 1 exclusive, we must have $s_1\hat{f} < s_2$ and $s_2\hat{f} < s_1$.

In the following treatment, we assume, without losing generality, that

$$s_1 \geqslant s_2 > 0. \qquad \text{(A4.15)}$$

Therefore, the condition $0 < \hat{x}_1 < 1$ is equivalent to

$$\hat{f} < \frac{s_2}{s_1}, \tag{A4.16}$$

or, using (A4.13),

$$\sqrt{D} > 2(1 - i_s) - (1 - 2i_s)S - 4i_s\left(\frac{s_2}{s_1}\right). \tag{A4.17}$$

Let us now investigate the stability of the equilibrium given by (A4.8) and (A4.13). Since we know that in this case there is only one non-trivial equilibrium point between 0 and 1, it is possible to examine the stability by investigating the change of gene frequencies at the neighborhood of 0 and 1. First, at the neighborhood of $x_1 = 1$ where the frequency of $A_1$ is very high, it can be easily seen that $A_1$ tends to decrease in frequency, because gene $A_2$, either in predominantly homozygous state or in combination with $A_1$, is always fitter than $A_1$ in this region.

Next, consider the neighborhood of $x_1 = 0$. In this region the frequency of $A_2$ is very high with the result that most individuals in the population are $A_2A_2$. Thus the rate of change in the frequencies of individuals having gene $A_1$ may be expressed by the following set of equations derived from (A4.1) neglecting higher-order terms:

$$\begin{aligned} P_{11}' &= \frac{1 - s_1}{1 - s_2} SP_{11} + \frac{1}{1 - s_2}\frac{S}{2} P_{12} \\ P_{12}' &= \frac{1 - s_1}{1 - s_2} RP_{11} + \frac{1}{1 - s_2}\left(R + \frac{S}{2}\right) P_{12}. \end{aligned} \tag{A4.18}$$

Here we neglect the trivial case of $s_1 = s_2 = 1$. The characteristic equation of the above transformation is

$$\begin{vmatrix} \dfrac{1 - s_1}{1 - s_2} S - \lambda & \dfrac{1}{1 - s_2}\dfrac{S}{2} \\ \dfrac{1 - s_1}{1 - s_2} R & \dfrac{1}{1 - s_2}\left(R + \dfrac{S}{2}\right) - \lambda \end{vmatrix} = 0,$$

from which we obtain two characteristic roots

$$\lambda_1 = \frac{1 + (1/2 - s_1)S + \sqrt{1 - S + S^2(s_1 - 1/2)^2}}{2(1 - s_2)}$$

$$\lambda_2 = \frac{1 + (1/2 - s_1)S - \sqrt{1 - S + S^2(s_1 - 1/2)^2}}{2(1 - s_2)} \tag{A4.19}$$

both of which are positive. Since $\lambda_1$ is larger than $\lambda_2$, we may write, for large $t$,

$$P_{11}^{(t)} \sim C_1\lambda_1^t, \qquad P_{12}^{(t)} \sim C_2\lambda_2^t,$$

where $t$ stands for time in generations and $C_1$, $C_2$ are positive constants. The necessary and sufficient condition for the equilibrium to be stable is that the frequency of $A_1$, i.e.

$$x_1^{(t)} = P_{11}^{(t)} + P_{12}^{(t)} \sim (C_1 + C_2)\lambda_1^t$$

tends to increase. This is equivalent to saying that $\lambda_1$ must be larger than unity, i.e.

$$\frac{1 + (1/2 - s_1)S + \sqrt{1 - S + S^2(s_1 - 1/2)^2}}{2(1 - s_2)} > 1,$$

which leads to

$$\sqrt{1 - S + S^2(s_1 - 1/2)^2} > (1 - 2s_2) - (1/2 - s_1)S \tag{A4.20}$$

as the condition for stability. It may be convenient to classify cases satisfying the above inequality into the following three classes.

(i) If

$$(1 - 2s_2) - (1/2 - s_1)S < 0,$$

we have

$$2(1 - 2s_2) < (1 - 2s_1)S.$$

Since $0 \leqq S \leqq 1$ and $(1 - 2s_2) > (1 - 2s_1)$, this is equivalent to the condition:

$$s_1, s_2 > 1/2,$$

namely, the heterozygote is more than twice as fit as either homozygote. In this case equilibrium is stable irrespective of the amount of self-fertilization.

(ii) If

$$(1 - 2s_2) - (1/2 - s_1)S = 0,$$

we have, noting that $(1 - 2s_2) \geqq (1 - 2s_1)$,

$$2(1 - 2s_1) \leqq (1 - 2s_1)S,$$

from which we obtain, assuming $S \neq 1$,

$$s_1 = s_2 = 1/2.$$

In this case, (A4.20) reduces to

$$\sqrt{1 - S} > 0, \tag{A4.21}$$

and we conclude that the population must not be exclusively self-fertilizing for the equilibrium to be stable.

(iii) If

$$(1 - 2s_2) - (1/2 - s_1)S > 0$$

we have

$$2(1 - 2s_2) > (1 - 2s_1)S.$$

In this case $s_2$ must be smaller than $1/2$. Furthermore, by squaring each side of (A4.20) we obtain

$$S < \frac{2s_2(1 - s_2)}{s_1 + s_2 - 2s_1s_2}. \tag{A4.22}$$

Since the right side of this inequality can never exceed unity, this, together with (A4.21), means that unless the heterozygote is more than twice as fit as either homozygote, the amount of self-fertilization must be restricted to the extent shown by the inequality (A4.22) for the equilibrium to be stable. Note that (A4.22) reduces to (A4.21) if we put $s_1 = s_2 = 1/2$. Furthermore, we can show that if (A4.22) holds, the right-hand side of (A4.17) is positive, and therefore, squaring both sides of (A4.17), we obtain exactly the same condition as (A4.22).

As an application of this condition, we can show that a pair of alleles with $s_1 = 0.5$, $s_2 = 0.1$ can maintain balanced polymorphism only in a population in which the proportion of self-fertilization is less than 36%. In Figure A.1 the regions of $s_1$ and $s_2$ allowing a stable polymorphism are drawn for various values of $S$ in such a way that the triangular region above each curve of given value of $S$ is the region of stability. The main results of this section were published previously by Kimura (1960c) in Japanese.

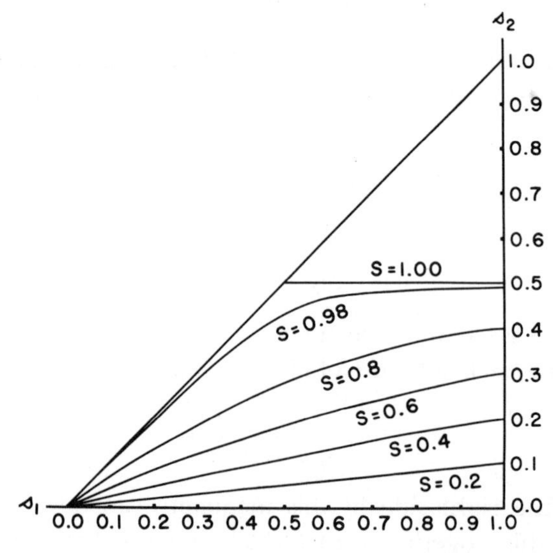


FIGURE A.1. The relation among $S$, $s_1$ and $s_2$ that gives a stable polymorphism due to overdominant alleles in a partially self-fertilizing population. For details see text. (From Kimura 1960c.)

# Bibliography


Bodmer, W. F., and J. Felsenstein. 1967. Linkage and selection: Theoretical analysis of the deterministic two locus random mating model. *Genetics* 57:237–265.

Bodmer, W. F., and P. A. Parsons. 1962. Linkage and recombination in evolution. *Advan. Genetics* 11:1–100.

Callan, H. G. 1967. The organization of genetic units in chromosomes. *J. Cell Sci.* 2:1–7.

Carr, R. N., and R. F. Nassar. 1970. Effects of selection and drift on the dynamics of finite populations. I. Ultimate probability of fixation of a favorable allele. *Biometrics* 26:41–49.

Cavalli-Sforza, L. L. 1958. Some data on the genetic structure of human populations. *Proc. X Int. Cong. Genetics* I:389–407.

Cavalli-Sforza, L. L. 1962. The distribution of migration distances: Models, and applications to genetics. In Entretiens de Monaco en Sciences Humains, Jean Sutter (ed.), Editions "Sciences humaines" Diffusion, Hachette. Pp. 139–158.

Cavalli-Sforza, L. L. 1969. "Genetic drift" in an Italian population. *Sci. Amer.* 221(No. 2):30–37.

Cavalli-Sforza, L. L., I. Barrai and A. W. F. Edwards. 1964. Analysis of human evolution under random genetic drift. *Cold Spring Harbor Symp. Quant. Biol.* 29:9–20.

Cavalli-Sforza, L. L., M. Kimura and I. Barrai. 1966. The probability of consanguineous marriages. *Genetics* 54: 37–60.

Chovnick, A. 1966. Genetic organization in higher organisms. *Proc. Roy. Soc., B.* 164:198–208.

Chovnick, A., A. Schalet, R. P. Kernaghan and J. Talsma. 1962. The resolving power of genetic fine structure analysis in higher organisms as exemplified by *Drosophila. Amer. Nat.* 96:281–296.

Clarke, B. 1969. The evidence for apostatic selection. *Heredity* 24:347–352.

Clarke, B., and P. O'Donald. 1964. Frequency-dependent selection. *Heredity* 19:201–206.

Cockerham, C. C. 1954. An extension of the concept of partitioning hereditary variance for analysis of covariances among relatives when epistasis is present. *Genetics* 39:859–882.

Cox, E. C., and Ch. Yanofsky. 1967. Altered base ratios in the DNA of an *Escherichia coli* mutator strain. *Proc. Nat. Acad. Sci.* 58:1895–1902.

Crow, J. F. 1954. Breeding structure of populations II. Effective population number. In Statistics and Mathematics in Biology, Kempthorne et al. (ed.), Iowa State College Press, Ames, Iowa. Pp. 543–556.

Crow, J. F. 1957. Possible consequences of an increased mutation rate. *Eugenics Quarterly* 4:67–80.

Crow, J. F. 1958. Some possibilities for measuring selection intensities in man. *Human Biol.* 30:1–13.

Crow, J. F. 1968a. The cost of evolution and genetic loads. In Haldane and Modern Biology, K. R. Dronamraju (ed.), Johns Hopkins Press, Baltimore. Pp. 165–178.

Crow, J. F. 1968b. Some analysis of hidden variability in *Drosophila* populations. In Population Biology and Evolution, R. Lewontin (ed.), Syracuse Univ. Press, New York. pp. 71–86.

Crow, J. F. 1969. Molecular genetics and population genetics. *Proc. XII Int. Cong. Genetics* 3:105–113.

Crow, J. F. 1970. Genetics loads and the cost of natural selection. In Biomathematics, Vol. 1, Mathematical Topics in Population Genetics. K. Kojima (ed.), Springer-Verlag, Berlin. Pp. 128–177.

Crow, J. F., and M. Kimura. 1963. The theory of genetic loads. *Proc. XI Int. Cong. Genetics* 3:495–506.

Crow, J. F., and M. Kimura. 1965. Evolution in sexual and asexual populations. *Amer. Nat.* 99:439–450.

Crow, J. F., and M. Kimura. 1969. Evolution in sexual and asexual populations: A reply. *Amer. Nat.* 103:89–91.

Crow, J. F., and M. Kimura. 1970. An Introduction to Population Genetics Theory. Harper and Row, New York.

Crow, J. F., and M. Kimura. 1971. The effective number of a population with overlapping generations: A correction and further discussion. *Amer. J. Human Genetics* (in press).

Crow, J. F., and A. P. Mange. 1965. Measurement of inbreeding from the frequency of marriages between persons of the same surname. *Eugenics Quarterly* 12:199–203.

Crow, J. F., and T. Maruyama. 1971. The number of neutral alleles maintained in a finite, geographically structured population. *Theor. Pop. Biol.* (in press).

Crow, J. F., and N. E. Morton. 1955. Measurement of gene frequency drift in small populations. *Evolution* 9:202–214.

Crow, J. F., and N. E. Morton. 1960. The genetic load due to mother-child incompatibility. *Amer. Nat.* 94:413–419.

Crow, J. F., and R. G. Temin. 1964. Evidence for the partial dominance of recessive lethal genes in natural populations of *Drosophila. Amer. Nat.* 98:21–33.

Dayhoff, M. O. 1969. Atlas of Protein Sequence and Structure 1969. National Biomedical Research Foundation, Silver Spring, Maryland.

Deevey, E. S., Jr. 1960. The human population. *Sci. Amer.* 203 (No. 3):195–204.

Dempster, E. R. 1955. Genetic models in relation to animal breeding problems. *Biometrics* 11:525–536.

Dobzhansky, Th. 1955. A review of some fundamental

concepts and problems of population genetics. *Cold Spring Harbor Symp.* 20:1–15.

Dunn, L. C. 1957. Evidence of evolutionary forces leading to the spread of lethal genes in wild populations of house mice. *Proc. Nat. Acad. Sci.* 43:158–163.

Ewens, W. J. 1963. Numerical results and diffusion approximations in a genetic process. *Biometrika* 50:241–249.

Ewens, W. J. 1968. A genetic model having complex linkage behaviour. *Theor. Appl. Genetics* 38:140–143.

Ewens, W. J. 1969. Population Genetics. Methuen, London.

Feller, W. 1967. On fitness and the cost of natural selection. *Genetical Res.* 9:1–15.

Felsenstein, J. 1969. The effective size of a population with overlapping generations. *Genetics* 61:s18.

Felsenstein, J. 1971. On the biological significance of the cost of gene substitution. *Amer. Nat.* 105:1–11.

Fincham, J. R. S. 1966. Genetic Complementation. Benjamin, New York.

Fisher, R. A. 1922. On the dominance ratio. *Proc. Roy. Soc. Edinburgh* 42:321–341.

Fisher, R. A. 1930a. The distribution of gene ratios for rare mutations. *Proc. Roy. Soc. Edinburgh* 50:205–220.

Fisher, R. A. 1930b. The Genetical Theory of Natural Selection. Clarendon Press, Oxford.

Fitch, W. M., and E. Markowitz. 1970. An improved method for determining codon variability in a gene and its application to the rate of fixation of mutations in evolution. *Biochemical Genetics* 4:579.

Ford, E. B. 1965. Genetic Polymorphism. Faber and Faber, London.

Franklin, I., and R. C. Lewontin. 1970. Is the gene the unit of selection? *Genetics* 65:707–734.

Freese, E. 1962. On the evolution of base composition of DNA. *J. Theor. Biol.* 3:82–101.

Frydenberg, O. 1963. Population studies of a lethal mutant

in *Drosophila melanogaster* I. Behaviour in populations with discrete generations. *Hereditas* 50:89–116.

Gibson, T. C., M. L. Scheppe and E. C. Cox. 1970. Fitness of an *Escherichia coli* mutator gene. *Science* 169:686–688.

Greenberg, Rayla, and J. F. Crow. 1960. A comparison of the effect of lethal and detrimental chromosomes from *Drosophila* populations. *Genetics* 45:1153–1168.

Gumbel, E. J. 1958. Statistics of Extremes. Columbia Univ. Press, New York.

Haga, T. 1969. Structure and dynamics of natural populations of a diploid *Trillium.* In Chromosomes Today Vol. 2, C. D. Darlington and K. R. Lewis (ed.), Oliver & Boyd, Edinburgh. Pp. 207–217.

Haga, T., and M. Kurabayashi. 1953. Genom and polyploidy in the genus *Trillium.* IV. Genom analysis by means of differential reaction of chromosome segment to low temperature. *Cytologia* 18:13–28.

Haldane, J. B. S. 1924a. A mathematical theory of natural and artificial selection. Part I. *Trans. Camb. Phil. Soc.* 23:19–41.

Haldane, J. B. S. 1924b. A mathematical theory of natural and artificial selection. II. The influence of partial self-fertilization, inbreeding, assortative mating, and selective fertilization on the composition of Mendelian populations, and on natural selection. *Proc. Camb. Phil. Soc.* 1:158–163.

Haldane, J. B. S. 1926. A mathematical theory of natural and artificial selection. Part III. *Proc. Camb. Phil. Soc.* 23:363–372.

Haldane, J. B. S. 1927a. A mathematical theory of natural and artificial selection. Part IV. *Proc. Camb. Phil. Soc.* 23:607–615.

Haldane, J. B. S. 1927b. A mathematical theory of natural and artificial selection. Part V. Selection and mutation. *Proc. Camb. Phil. Soc.* 23:838–844.

Haldane, J. B. S. 1930. A mathematical theory of natural

and artificial selection. Part VI. Isolation. *Proc. Camb. Phil. Soc.* 26:220–230.

Haldane, J. B. S. 1931a. A mathematical theory of natural and artificial selection. Part VII. Selection intensity as a function of mortality rate. *Proc. Camb. Phil. Soc.* 27:131–136.

Haldane, J. B. S. 1931b. A mathematical theory of natural and artificial selection. Part VIII. Metastable populations. *Proc. Camb. Phil. Soc.* 27:137–142.

Haldane, J. B. S. 1937. The effect of variation on fitness. *Amer. Nat.* 71:337–349.

Haldane, J. B. S. 1939. The spread of harmful autosomal recessive genes in human populations. *Ann. Eugenics* 9:232–237.

Haldane, J. B. S. 1949. Suggestions as to quantitative measurement of rates of evolution. *Evolution* 3:51–56.

Haldane, J. B. S. 1954. The statics of evolution. In Evolution as a Process, J. Huxley et al. (ed.), George Allen & Unwin, London. pp. 109–121.

Haldane, J. B. S. 1957. The cost of natural selection. *J. Genetics* 55:511–524.

Haldane, J. B. S. 1959. Personal communication to J. F. Crow.

Haldane, J. B. S. 1960. More precise expressions for the cost of natural selection. *J. Genetics* 57:351–360.

Harris, H. 1966. Enzyme polymorphism in man. *Proc. Roy. Soc., B.* 164:298–310.

Harris, H. 1969. Enzyme and protein polymorphism in human populations. *British Medical Bulletin* 25:5–13.

Hazel, L. N. 1943. The genetic basis for constructing selection indexes. *Genetics* 28:476–490.

Hill, W. G. 1969. On the theory of artificial selection in finite populations. *Genetical Res.* 13:143–163.

Hill, W. G., and A. Robertson. 1966. The effect of linkage on limits to artificial selection. *Genetical Res.* 8:269–294.

Hill, W. G., and A. Robertson. 1968. Linkage disequilibrium in finite populations. *Theor. Appl. Genetics* 38:226–231.

Huxley, J. 1955. Morphism and evolution. *Heredity* 9:1–52.

Karlin, S., and M. W. Feldman. 1969. Linkage and selection: New equilibrium properties of the two-locus symmetric viability model. *Proc. Nat. Acad. Sci.* 62:70–74.

Karlin, S., and M. W. Feldman. 1970. Linkage and selection: Two locus symmetric viability model. *Theor. Pop. Biol.* 1:39–71.

Karlin, S., and J. McGregor. 1968. Rates and probabilities of fixation for two locus random mating populations without selection. *Genetics* 58:141–159.

Kempthorne, O. 1957. An Introduction to Genetic Statistics. Wiley, New York.

Kerr, W. E. 1967. Multiple alleles and genetic load in bees. *J. Apicultural Res.* 6:61–64.

Kimura, M. 1953. "Stepping stone" model of population. *Ann. Rep. Nat. Inst. Genetics* 3:63–65.

Kimura, M. 1955. Stochastic processes and distribution of gene frequencies under natural selection. *Cold Spring Harbor Symp.* 20:33–53.

Kimura, M. 1956a. A model of a genetic system which leads to closer linkage by natural selection. *Evolution* 10:278–287.

Kimura, M. 1956b. Rules for testing stability of a selective polymorphism. *Proc. Nat. Acad. Sci.* 42:336–340.

Kimura, M. 1957. Some problems of stochastic processes in genetics. *Ann. Math. Stat.* 28:882–901.

Kimura, M. 1960a. Optimum mutation rate and degree of dominance as determined by the principle of minimum genetic load. *J. Genetics* 57:21–34.

Kimura, M. 1960b. Genetic load of a population and its significance in evolution. (Japanese with English summary) *Jap. J. Genetics* 35:7–33.

Kimura, M. 1960c. Outlines of Population Genetics. Baifukan, Tokyo (in Japanese).

Kimura, M. 1961a. Natural selection as the process of accumulating genetic information in adaptive evolution. *Genetical Res.* 2:127–140.

Kimura, M. 1961b. Some calculations on the mutational load. *Jap. J. Genetics* 36 Supplement:179–190.

Kimura, M. 1962. On the probability of fixation of mutant genes in a population. *Genetics* 47:713–719.

Kimura, M. 1964. Diffusion models in population genetics. *J. Appl. Prob.* 1:177–232.

Kimura, M. 1965a. Attainment of quasi-linkage equilibrium when gene frequencies are changing by natural selection. *Genetics* 52:875–890.

Kimura, M. 1965b. A stochastic model concerning the maintenance of genetic variability in quantitative characters. *Proc. Nat. Acad. Sci.* 54:731–736.

Kimura, M. 1966. Two loci polymorphism as a stationary point. *Ann. Rep. Nat. Inst. Genetics* 17:65–67.

Kimura, M. 1967. On the evolutionary adjustment of spontaneous mutation rates. *Genetical Res.* 9:23–34.

Kimura, M. 1968a. Evolutionary rate at the molecular level. *Nature* 217:624–626.

Kimura, M. 1968b. Genetic variability maintained in a finite population due to mutational production of neutral and nearly neutral isoalleles. *Genetical Res.* 11: 247–269.

Kimura, M. 1969a. The rate of molecular evolution considered from the standpoint of population genetics. *Proc. Nat. Acad. Sci.* 63:1181–1188.

Kimura, M. 1969b. The number of heterozygous nucleotide sites maintained in a finite population due to steady flux of mutations. *Genetics* 61:893–903.

Kimura, M. 1970a. The length of time required for a selectively neutral mutant to reach fixation through random frequency drift in a finite population. *Genetical Res.* 15:131–133.

Kimura, M. 1970b. Stochastic processes in population genetics, with special reference to distribution of gene frequencies and probability of gene fixation. In Biomathematics, Vol. 1, Mathematical Topics in Population Genetics. K. Kojima (ed.), Springer-Verlag, Berlin. Pp. 178–209.

Kimura, M., and J. F. Crow. 1963a. The measurement of effective population number. *Evolution* 17:279–288.

Kimura, M., and J. F. Crow. 1963b. On the maximum avoidance of inbreeding. *Genetical Res.* 4:399–415.

Kimura, M., and J. F. Crow. 1964. The number of alleles that can be maintained in a finite population. *Genetics* 49:725–738.

Kimura, M., and J. F. Crow. 1969. Natural selection and gene substitution. *Genetical Res.* 13:127–141.

Kimura, M., and H. Kayano. 1961. The maintenance of supernumerary chromosomes in wild populations of *Lilium callosum* by preferential segregation. *Genetics* 46:1699–1712.

Kimura, M., and T. Maruyama. 1966. The mutational load with epistatic gene interactions in fitness. *Genetics* 54: 1337–1351.

Kimura, M., and T. Maruyama. 1969. The substitutional load in a finite population. *Heredity* 24:101–114.

Kimura, M., and T. Maruyama. 1971. Pattern of neutral polymorphism in a geographically structured population. *Genetical Res.* (in press).

Kimura, M., T. Maruyama and J. F. Crow. 1963. The mutation load in small populations. *Genetics* 48:1303–1312.

Kimura, M., and T. Ohta. 1969a. The average number of generations until fixation of a mutant gene in a finite population. *Genetics* 61:763–771.

Kimura, M., and T. Ohta. 1969b. The average number of generations until extinction of an individual mutant gene in a finite population. *Genetics* 63:701–709.

Kimura, M., and T. Ohta. 1970a. Genetic load at a polymorphic locus which is maintained by frequency-dependent selection. *Genetical Res.* 16:145–150.

Kimura, M., and T. Ohta. 1970b. Probability of fixation of a mutant gene in a finite population when selective advantage decreases with time. *Genetics* 65:525–534.

Kimura, M., and T. Ohta. 1971. Protein polymorphism as a phase of molecular evolution. *Nature* 229:467–469.

Kimura, M., and G. H. Weiss. 1964. The stepping stone model of population structure and the decrease of genetic correlation with distance. *Genetics* 49:561–576.

King, J. L. 1967. Continuously distributed factors affecting fitness. *Genetics* 55:483–492.

King, J. L., and T. H. Jukes. 1969. Non-Darwinian evolution: Random fixation of selectively neutral mutations. *Science* 164:788–798.

Kojima, K., and T. M. Kelleher. 1962. Survival of mutant genes. *Amer. Nat.* 96:329–346.

Kojima, K., and K. M. Yarbrough. 1967. Frequency-dependent selection at the esterase 6 locus in *Drosophila melanogaster. Proc. Nat. Acad. Sci.* 57:645–649.

Laird, C. D., B. L. McConaughy and B. J. McCarthy. 1969. Rate of fixation of nucleotide substitutions in evolution. *Nature* 224:149–154.

Latter, B. D. H., and A. Robertson. 1962. The effects of inbreeding and artificial selection on reproductive fitness. *Genetical Res.* 3:110–138.

Lewontin, R. C. 1967. An estimate of average heterozygosity in man. *Amer. J. Human Genetics* 19:681–685.

Lewontin, R. C., and J. L. Hubby. 1966. A molecular approach to the study of genic heterozygosity in natural populations. II. Amount of variation and degree of heterozygosity in natural populations of *Drosophila pseudoobscura*. *Genetics* 54:595–609.

Lewontin, R. C., and K. Kojima. 1960. The evolutionary dynamics of complex polymorphisms. *Evolution* 14:458–472.

Li, C. C. 1967. Genetic equilibrium under selection. *Biometrics* 23:397–484.

Lifschytz, E., and R. Falk. 1969. Fine structure analysis of a chromosome segment in *Drosophila melanogaster*. Analysis of ethyl methansulphonate-induced lethals. *Mut. Res.* 8:147–155.

Malécot, G. 1948. Les Mathématiques de l'Hérédité. Masson et Cie., Paris.

Malécot, G. 1955. Decrease of relationship with distance. *Cold Spring Harbor Symp.* 20:52–53.

Malécot, G. 1959. Les modéles stochastiques en génétique de population. *Publications de l'Institut de Statistique de l'Université de Paris* 8:173–210.

Malécot, G. 1967. Identical loci and relationship. *Proc. Fifth Berkeley Symp. Math. Stat. Prob.* IV:317–332 (Univ. of California Press, Berkeley).

Marshall, D. R., and R. W. Allard. 1971. Maintenance of isozyme polymorphisms in natural populations of *Avena barbata*. *Genetical Res.* (in press).

Maruyama, T. 1969. Genetic correlation in the stepping stone model with non-symmetrical migration rates. *J. Appl. Prob.* 6:463–477.

Maruyama, T. 1970a. On the rate of decrease of heterozygosity in circular stepping stone models of populations. *Theor. Pop. Biol.* 1:101–119.

Maruyama, T. 1970b. Rate of decrease of genetic variability in a subdivided population. *Biometrika* (submitted).

Maruyama, T. 1970c. Effective number of alleles in subdivided population. *Theor. Pop. Biol.* 1:273–306.

Maruyama, T. 1971a. Analysis of population structure II.

Two-dimensional stepping stone models of finite length and other geographically structured populations. *Ann. Human Genet.* (in press).

Maruyama, T. 1971b. Speed of gene substitution in a geographically structured population. *Amer. Nat.* (in press).

Maynard Smith, J. 1968a. "Haldane's dilemma" and the rate of evolution. *Nature* 219:1114–1116.

Maynard Smith, J. 1968b. Evolution in sexual and asexual populations. *Amer. Nat.* 102:469–473.

Milkman, R. D. 1967. Heterosis as a major cause of heterozygosity in nature. *Genetics* 55:493–495.

Moran, P. A. P. 1964. On the nonexistence of adaptive topographies. *Ann. Human Genetics* 27:383–393.

Morton, N. E. 1960. The mutational load due to detrimental genes in man. *Amer. J. Human Genetics* 12:348–364.

Morton, N. E., J. F. Crow and H. J. Muller. 1956. An estimate of the mutational damage in man from data on consanguineous marriages. *Proc. Nat. Acad. Sci.* 42: 855–863.

Mukai, T. 1964. The genetic structure of natural populations of *Drosophila melanogaster.* I. Spontaneous mutation rate of polygenes controlling viability. *Genetics* 50:1–19.

Mukai, T. 1968. Experimental studies on the mechanism involved in the maintenance of genetic variability in *Drosophila* populations. (Japanese with English summary) *Jap. J. Genetics* 43:399–413.

Mukai, T. 1969a. The genetic structure of natural populations of *Drosophila melanogaster.* VII. Synergistic interaction of spontaneous mutant polygenes controlling viability. *Genetics* 61:749–761.

Mukai, T. 1969b. Maintenance of polygenic and isoallelic variation in populations. *Proc. XII Int. Cong. Genetics* 3:293–308.

Muller, H. J. 1932. Some genetic aspects of sex. *Amer. Nat.* 66:118–138.

Muller, H. J. 1950. Our load of mutations. *Amer. J. Human Genetics* 2:111–176.

Muller, H. J. 1958. Evolution by mutation. *Bull. Amer. Math. Soc.* 64:137–160.

Muller, H. J. 1960. Evolution and genetics. *Academia Nazionale dei Lincei* 47:15–37.

Muller, H. J. 1966. The gene material as the initiator and the organizing basis of life. *Amer. Nat.* 100:493–517.

Neel, J. V., and W. J. Schull. 1962. The effect of inbreeding on mortality and morbidity in two Japanese cities. *Proc. Nat. Acad. Sci.* 48:573–582.

Nei, M. 1965. Variation and covariation of gene frequencies in subdivided populations. *Evolution* 19:256–258.

Nei, M. 1967. Modification of linkage intensity by natural selection. *Genetics* 57:625–641.

Nei, M. 1968a. The frequency distribution of lethal chromosomes in finite populations. *Proc. Nat. Acad. Sci.* 60: 517–524.

Nei, M. 1968b. Evolutionary change of linkage intensity. *Nature* 218:1160–1161.

Nei, M. 1970. Fertility excess necessary for gene substitution in regulated populations. *Genetics* (in press).

Nei, M., and Y. Imaizumi. 1966a. Genetic structure of human populations I. Local differentiation of blood group gene frequencies in Japan. *Heredity* 21:9–35.

Nei, M., and Y. Imaizumi. 1966b. Genetic structure of human populations II. Differentiation of blood group gene frequencies among isolated human populations. *Heredity* 21:183–190.

Nei, M., and M. Murata. 1966. Effective population size when fertility is inherited. *Genetical Res.* 8:257–260.

O'Brien, S. J., and R. J. MacIntyre. 1969. An analysis of gene-enzyme variability in natural populations of

*Drosophila melanogaster* and *D. simulans*. *Amer. Nat.* 103: 97–113.

Ohno, S. 1970. Evolution by Gene Duplication. Springer-Verlag, Berlin.

Ohta, T. 1971. Associative overdominance caused by linked detrimental mutations. Submitted to *Genet. Res.*

Ohta, T., and M. Kimura. 1969a. Linkage disequilibrium due to random genetic drift. *Genetical Res.* 13:47–55.

Ohta, T., and M. Kimura. 1969b. Linkage disequilibrium at steady state determined by random genetic drift and recurrent mutation. *Genetics* 63:229–238.

Ohta, T., and M. Kimura. 1970. Development of associative overdominance through linkage disequilibrium in finite populations. *Genetical Res.* 16:165–177.

Ohta, T., and M. Kimura. 1971a. Linkage disequilibrium between two segregating nucleotide sites under steady flux of mutations in a finite population. *Genetics* (in press).

Ohta, T., and M. Kimura. 1971b. Behavior of neutral mutants influenced by associated overdominant loci in finite populations. *Genetics* (submitted).

Ohta, T., and M. Kimura. 1971c. Functional organization of genetic material as a product of molecular evolution. *Nature* (in press).

Prakash, S., R. C. Lewontin and J. L. Hubby. 1969. A molecular approach to the study of genic heterozygosity in natural populations. IV. Patterns of genic variation in central, marginal and isolated populations of *Drosophila pseudoobscura*. *Genetics* 61:841–858.

Robertson, A. 1960. A theory of limits in artificial selection. *Proc. Roy. Soc.* 153(B):234–249.

Robertson, A. 1961. Inbreeding in artificial selection programmes. *Genetical Res.* 2:189–194.

Robertson, A. 1962. Selection for heterozygotes in small populations. *Genetics* 47:1291–1300.

Robertson, A. 1964. The effect of non-random mating within inbred lines on the rate of inbreeding. *Genetical Res.* 5:164–167.

Robertson, A. 1967. The nature of quantitative genetic variation. In Heritage from Mendel, R. A. Brink (ed.), Univ. Wisconsin Press, Madison. Pp. 265–280.

Robertson, A. 1968. The spectrum of genetic variation. In Population Biology and Evolution, R. Lewontin (ed.), Syracuse Univ. Press, New York. Pp. 5–16.

Robertson, A. 1970a. The reduction in fitness from genetic drift at heterotic loci in small populations. *Genetical Res.* 15:257–259.

Robertson, A. 1970b. A theory of limits in artificial selection with many linked loci. In Biomathematics, Vol. 1, Mathematical Topics in Population Genetics, K. Kojima (ed.), Springer-Verlag, Berlin. Pp. 246–288.

Schalet, A. 1969. Exchanges at the bobbed locus of *Drosophila melanogaster. Genetics* 63:133–153.

Schull, W. J., and J. V. Neel. 1965. The Effects of Inbreeding on Japanese Children. Harper and Row, New York.

Selander, R. K., W. G. Hunt and S. Y. Yang. 1969. Protein polymorphism and genic heterozygosity in two European subspecies of the house mouse. *Evolution* 23:379–390.

Selander, R. K., S. Y. Yang, R. C. Lewontin and W. E. Johnson. 1970. Genetic variation in the horseshoe crab (*Limulus polyphemus*), a phylogenetic "relic." *Evolution* 24:402–417.

Shaw, C. R. 1965. Electrophoretic variation in enzymes. *Science* 149:936–943.

Sheppard, P. M. 1969. Evolutionary genetics of animal populations: The study of natural populations. *Proc. XII Int. Cong. Genetics* 3:261–279.

Shinoda, T. 1967. Multiple molecular forms of xanthine dehydrogenase in *Drosophila. Ann. Rep. Nat. Inst. Genetics* No. 17, pp. 53–54.

Smith, H. F. 1936. A discriminant function for plant selection. *Ann. Eugenics* 7:240–250.

Stevenson, A. C. 1958. Some data, estimates and reflections on congenital and hereditary anomalies in the population of Northern Ireland. United Nations Document.

Sueoka, N. 1961. Variation and heterogeneity of base composition of deoxyribonucleic acids: A compilation of old and new data. *J. Mol. Biol.* 3:31–40.

Sueoka, N. 1962. On the genetic basis of variation and heterogeneity of DNA base composition. *Proc. Nat. Acad. Sci.* 48:582–592.

Sved, J. A., T. E. Reed and W. F. Bodmer. 1967. The number of balanced polymorphisms that can be maintained in a natural population. *Genetics* 55:469–481.

Tax, S., and C. Callender (eds.). 1959. Issues in Evolution. Vol. 3 of Evolution after Darwin. Univ. Chicago Press, Chicago.

Temin, R. G., H. U. Meyer, P. S. Dawson and J. F. Crow. 1969. The influence of epistasis on homozygous viability depression in *Drosophila melanogaster*. *Genetics* 61:497–519.

Vogel, F. 1964. A preliminary estimate of the number of human genes. *Nature* 201:847.

Wagner, M. 1868. Die Darwin'sche Theorie und das migrations-gesetz der Organismen. Ducken and Humbolt, Leipzig.

Wallace, B. 1950. Allelism of second chromosome lethals in *D. melanogaster*. *Proc. Nat. Acad. Sci.* 36:654–657.

Weiss, G. H., and M. Kimura. 1965. A mathematical analysis of the stepping stone model of genetic correlation. *J. Appl. Prob.* 2:129–149.

Wright, S. 1921. Systems of mating. II. The effects of inbreeding on the genetic composition of a population. *Genetics* 6:124–143.

Wright, S. 1931. Evolution in Mendelian populations. *Genetics* 16:97–159.

Wright, S. 1938a. The distribution of gene frequencies under irreversible mutation. *Proc. Nat. Acad. Sci.* 24:253–259.

Wright, S. 1938b. Size of population and breeding structure in relation to evolution. *Science* 87:430–431.

Wright, S. 1939. Statistical genetics in relation to evolution. In Actualités Scientifiques et Industrielles. No. 802. Exposés de Biométrie et de la Statistique Biologique. Hermann & Cie., Paris. pp. 5–64.

Wright, S. 1940. Breeding structure of populations in relation to speciation. *Amer. Nat.* 74:232–248.

Wright, S. 1943. Isolation by distance. *Genetics* 28:114–138.

Wright, S. 1945a. The differential equation of the distribution of gene frequencies. *Proc. Nat. Acad. Sci.* 31:382–389.

Wright, S. 1945b. Tempo and mode in evolution: A critical review. *Ecology* 26:415–419.

Wright, S. 1949. Adaptation and selection. In Genetics, Paleontology, and Evolution, G. Jepsen, et al. (ed.), Princeton Univ. Press, Princeton. Pp. 365–389.

Wright, S. 1951. The genetical structure of populations. *Ann. Eugenics* 15:323–354.

Wright, S. 1965. The interpretation of population structure by F-statistics with special regard to systems of mating. *Evolution* 19:395–420.

Wright, S. 1966. Polyallelic random drift in relation to evolution. *Proc. Nat. Acad. Sci. U.S.A.* 55:1074–1081.

Wright, S. 1967. "Surfaces" of selective value. *Proc. Nat. Acad. Sci.* 58:165–172.

Wright, S. 1969. Evolution and the Genetics of Populations. Vol. II. The theory of gene frequencies. Univ. Chicago Press, Chicago and London.

Wright, S., and Th. Dobzhansky. 1946. Genetics of natural populations. XII. Experimental reproduction of some of the changes caused by natural selection in certain populations of *Drosophila pseudoobscura. Genetics* 31:125–156.

Yamaguchi, M., T. Yanase, H. Nagano and N. Nakamoto. 1970. Effects of inbreeding on mortality in Fukuoka population. *Amer. J. Human Genetics* 22:145–159.

Yasuda, N. 1968. Distribution of matrimonial distance in the Mishima district. *Proc. XII Int. Cong. Genetics* 2:178–179.

Zuckerkandl, E., and L. Pauling. 1965. Evolutionary divergence and convergence in proteins. In Evolving Genes and Proteins, V. Bryson and H. J. Vogel (ed.), Academic Press, New York. Pp. 97–166.

Report of the United Nations Scientific Committee on the Effects of Atomic Radiation, 1958, New York.

# Author Index

# Subject Index